Hawker
HURRICANE
PORTFOLIO

Compiled by
R.M. Clarke

ISBN 0 948 207 922

Published by Brooklands Books in conjunction with

A BROOKLANDS AIRCRAFT PORTFOLIO

Portfolios in preparation will cover: Spitfire, Lancaster, Wellington, P51 Mustang etc.
Cover Photography by Stuart Howe

DISTRIBUTED BY

Motorbooks International
Osceola
Wisconsin 24020
USA

Brooklands Book Distribution Ltd
Holmerise, Sevenhills Road
Cobham, Surrey KT11 1ES
England

A BROOKLANDS AIRCRAFT PORTFOLIO

A BROOKLANDS AIRCRAFT PORTFOLIO

This is the first of a new series of books covering classic World War Two aircraft, compiled from contemporary material originally published in Flight, The Aeroplane *and* Aircraft Production. *We have combed through wartime volumes of these journals and selected representative features on these famous fighting aircraft, covering every aspect of each aeroplane: genealogies, technical appraisals, handling characteristics, combat and operational reports and so forth. In addition to a wealth of photographs each book contains one or more contemporary cutaway drawings, for which both* Flight *and* The Aeroplane *were noted. These articles have not been edited in any way and are straight, high-quality reprints from the original issues. In addition to this wealth of contemporary material further features have been reprinted from* Aeroplane Monthly, *successor to* The Aeroplane, *written with hindsight by pilots, engineers and crewmen who knew their aircraft intimately, inside and out.*

Whether or not you flew in these aircraft, this series conveys, in a unique form, the character and background of aircraft that fought for peace during six long years of war.

If you enjoy this book you will surely enjoy others in this series.

Richard Riding
Editor
Aeroplane Monthly

Printed in Hong Kong

ROYAL APPROVAL : H.M. King George VI inspects one of the Hawker Hurricanes at Northolt. With His Majesty is Sqn. Ldr. Gillan, who flew a Hurricane from Edinburgh to Northolt at more than 400 m.p.h. The general view below shows the clean lines of the Hurricane.

THE HURRICANE UNVEILED

A Detailed Description of the Fastest Fighter in Service in the World : Latest Aerodynamic Design Combined with Well-tried Construction

WHEN Mr. Sydney Camm and his staff began to plan their new eight-gun single-seater fighter, they had to make one very important decision before the design work could commence: Should they "go the whole hog," so to speak, and design a stressed-skin machine, of which they had at that time no experience, or should they try to marry the latest ideas in aerodynamic design to the type of structure with which they were familiar and of which they had years of successful experience? Many factors and considerations had to be taken into account. On the one hand, would the production of jigs and tools for stressed-skin construction, having to proceed parallel with the design of the machine, lead to such delays that deliveries by a certain date became problematic? On the other, if the older type of construction were adopted, could they be sure that the fabric covering would stand up to the speeds contemplated?

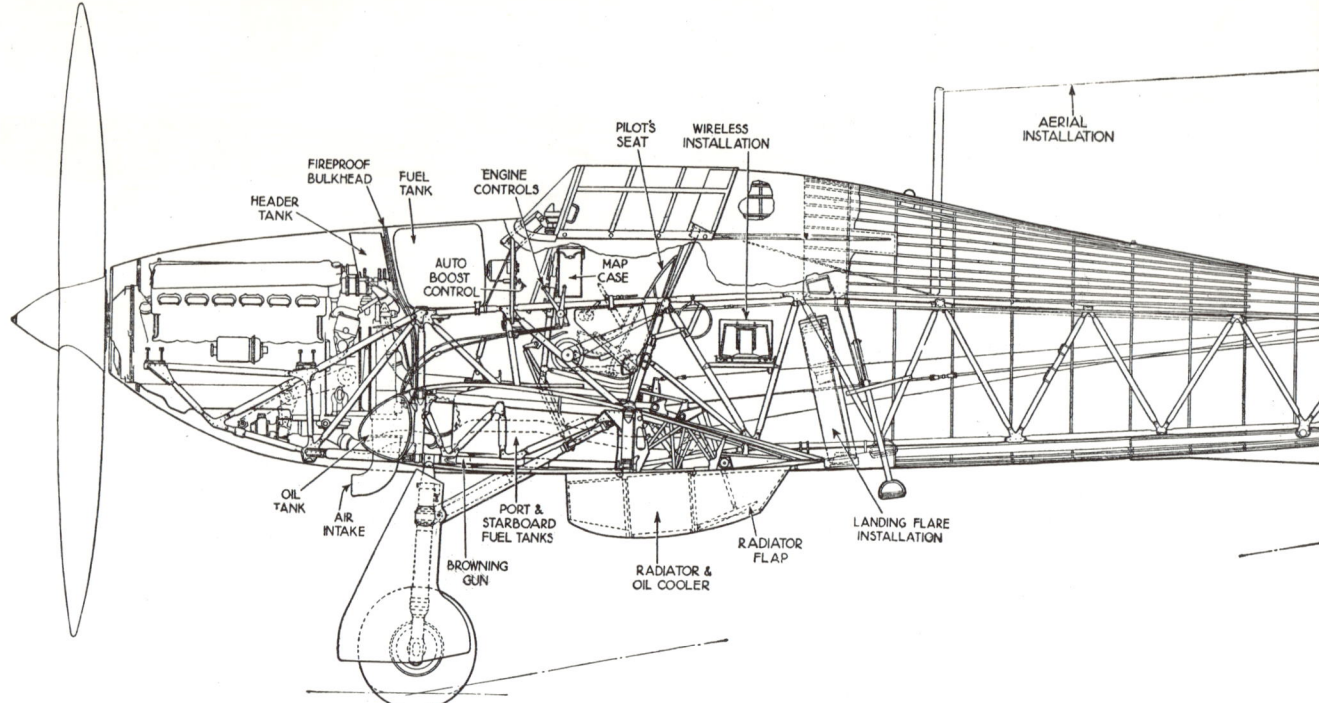

In the side elevation above but a small percentage of the equipment of the Hurricane can be shown. For instance, the eight machine-guns are placed in the wings, four on each side, and operated by remote control.

There were scores of other problems, but these were fundamental. In the end the familiar type of construction won, and the justification for the decision is to be found in the fact that the Hurricane is now in service with several R.A.F. squadrons, and that so far the fabric covering has not given any trouble. The performance is up to estimates, and, best of all, the young pilots who have to fly the machines have found no difficulty in handling them after a brief training on Miles Magisters. That the machines have to be treated with respect is but natural. For instance, an aeroplane as clean as the Hurricane picks up speed very quickly, and it does not do for the pilot to go wool-gathering and to let the machine get into a dive without him realising it. Otherwise he is apt to lose a lot of height before he is aware of the fact, and to get up to very high diving speeds. This is, of course, more particularly risky when the altitude is not great. But provided he remembers that he cannot stand the machine on its nose for many seconds, there is nothing in the handling of the Hurricane to worry the younger generation of pilots.

General Layout

The external appearance of the Hurricane is already well known to readers of *Flight* from the photographs published from time to time, and particularly the double-page picture in last week's issue. The machine is a low-wing cantilever monoplane with retractable undercarriage, a cabin roof over the pilot's cockpit, and a Rolls-Royce Merlin II liquid-cooled engine beautifully cowled.

Intended as it is for day and night flying, the Hurricane carries a very extensive military equipment, and its armament consists of no fewer than eight machine guns, which are housed inside the wings, four on each side. Fuel for something like two hours at full speed of the Merlin engine amounts to a good deal, and the ammunition for the machine guns is no small item either, so that, for a single-seater fighter the Hurricane is neither very small nor very light. Its loaded weight is in the neighbourhood of 6,000 lb., and the wing span is about 40ft.

It has already been mentioned that structurally the Hurricane follows those general principles which have been such a successful feature of the long series of Hawker machines which began with the Hart and all its variants and were employed in the Fury biplane, the fastest of the Hawker family until the advent of the Hurricane.

The "motif" of the wing construction is well brought out in the sketch on the right. The drag members produce, with the spars, a structure remarkably stiff in torsion.

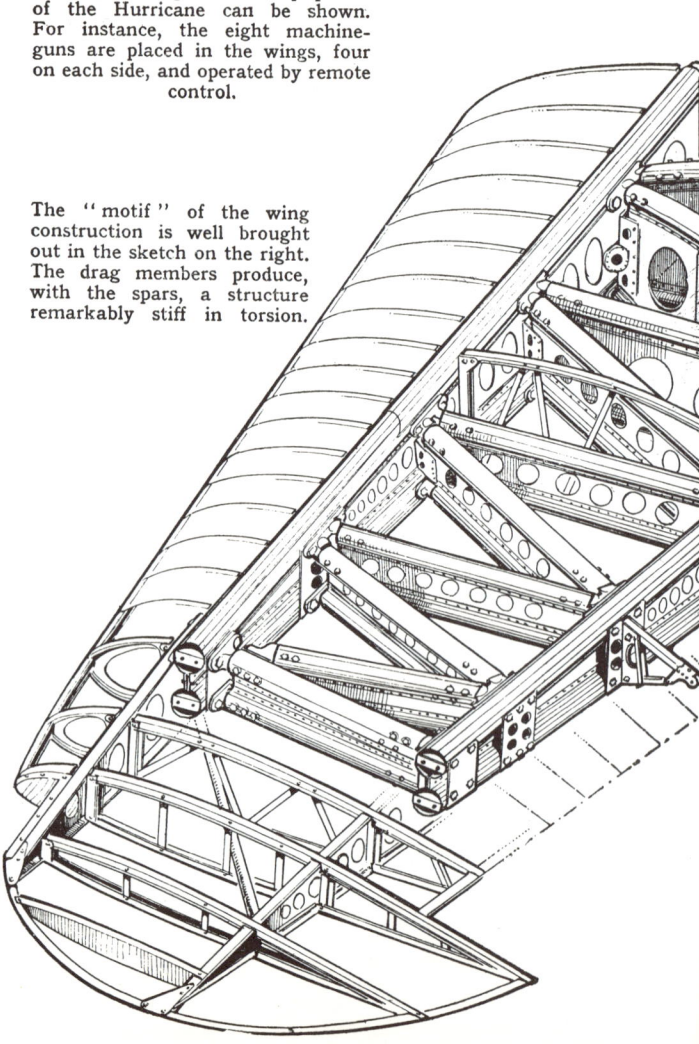

Steel and duralumin take their place side by side in the Hurricane, each according to its suitability for a particular purpose. In the fuselage, for example, the longerons are steel tubes, but the diagonal struts which form the bracing in the sides are of duralumin in the rear portion.

Little need be said of the fuselage construction, which has been familiar for many years, beyond recalling that use is made of circular-section tubes for the longerons, this section being turned into a square section with rounded corners at the points where the struts are attached by flat plates and bolts or tubular rivets. The struts run zig-zag fashion between top and bottom longerons, so that there is no wire bracing, but the struts in the top and bottom panels run transversely, and bracing is by streamline tie-rods.

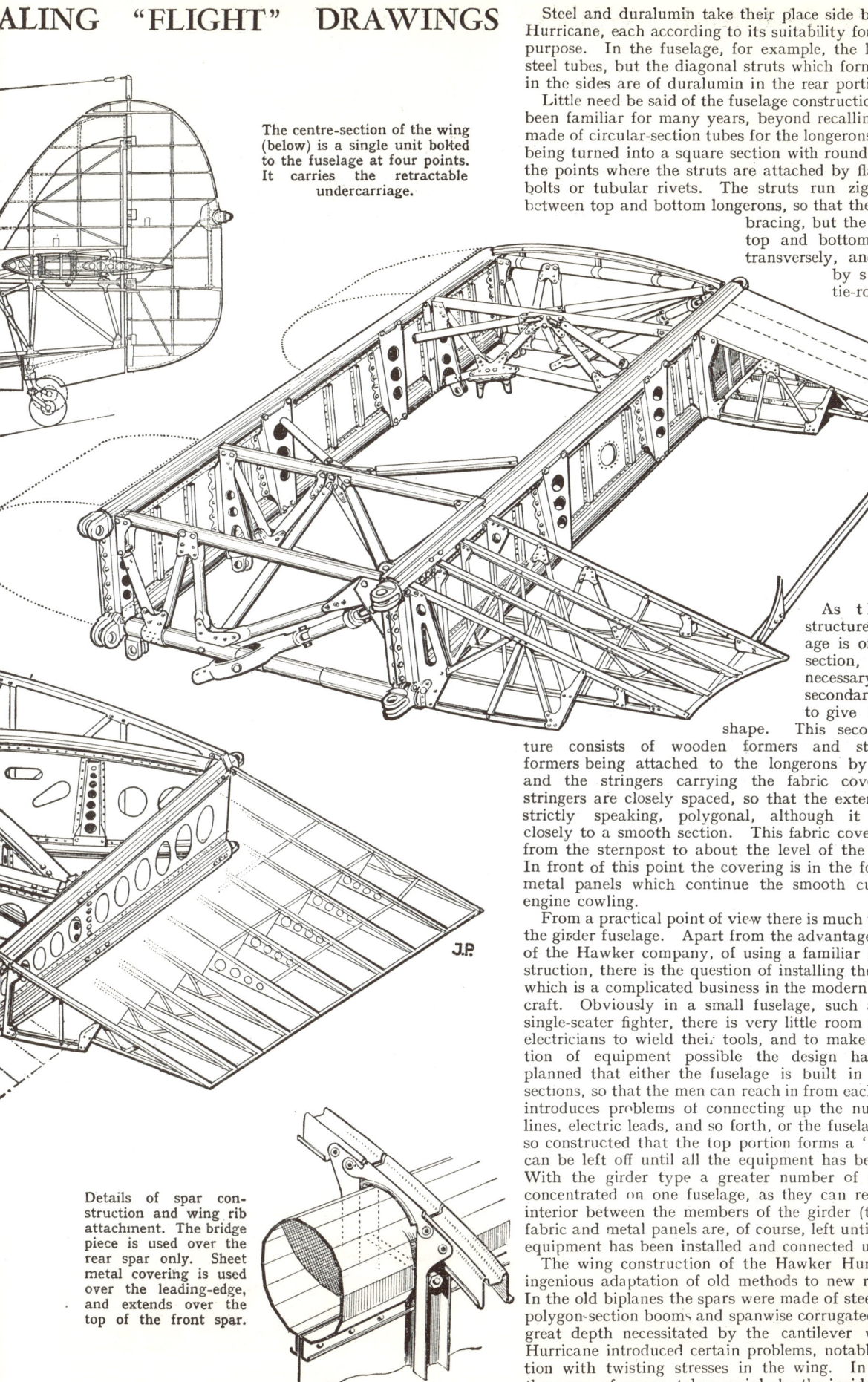

The centre-section of the wing (below) is a single unit bolted to the fuselage at four points. It carries the retractable undercarriage.

J.P.

Details of spar construction and wing rib attachment. The bridge piece is used over the rear spar only. Sheet metal covering is used over the leading-edge, and extends over the top of the front spar.

As the primary structure of the fuselage is of rectangular section, it has been necessary to add a secondary structure to give the rounded shape. This secondary structure consists of wooden formers and stringers, the formers being attached to the longerons by metal clips, and the stringers carrying the fabric covering. The stringers are closely spaced, so that the external form is, strictly speaking, polygonal, although it approaches closely to a smooth section. This fabric covering extends from the sternpost to about the level of the pilot's seat. In front of this point the covering is in the form of light metal panels which continue the smooth curves of the engine cowling.

From a practical point of view there is much to be said for the girder fuselage. Apart from the advantage, in the case of the Hawker company, of using a familiar form of construction, there is the question of installing the equipment, which is a complicated business in the modern military aircraft. Obviously in a small fuselage, such as that of a single-seater fighter, there is very little room for fitters or electricians to wield their tools, and to make the installation of equipment possible the design has to be so planned that either the fuselage is built in longitudinal sections, so that the men can reach in from each end, which introduces problems of connecting up the numerous pipe lines, electric leads, and so forth, or the fuselage has to be so constructed that the top portion forms a "lid," which can be left off until all the equipment has been installed. With the girder type a greater number of men can be concentrated on one fuselage, as they can reach into the interior between the members of the girder (the stringers, fabric and metal panels are, of course, left until most of the equipment has been installed and connected up).

The wing construction of the Hawker Hurricane is an ingenious adaptation of old methods to new requirements. In the old biplanes the spars were made of steel strip, with polygon-section booms and spanwise corrugated webs. The great depth necessitated by the cantilever wing of the Hurricane introduced certain problems, notably in connection with twisting stresses in the wing. In the biplane these are, of course, taken mainly by the incidence bracing.

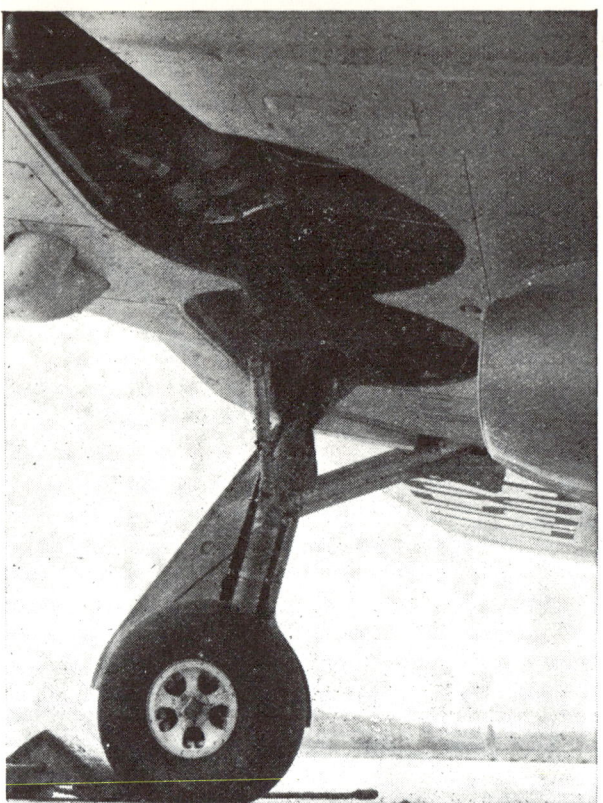

Why aircraft production is a slow business : This view of the pilot's cockpit gives some idea of the complexity of the equipment of a modern fighter.

The retractable undercarriage is raised laterally and inwardly, and the fairings cover the openings completely, leaving a smooth surface.

The manner in which Mr. Camm solved the problem is very interesting.

In a general way the spars of the Hurricane resemble those of the earlier biplanes. That is to say, they have

bulbous steel booms of polygonal section (much larger, of course, than those of the biplanes) at top and bottom, and a central web of flat sheet is riveted into the inwardly pointing open sides of the booms. In the centre section,

The Hurricane fuselage in skeleton. The construction is of the type used by the Hawker Company with great success for many years.

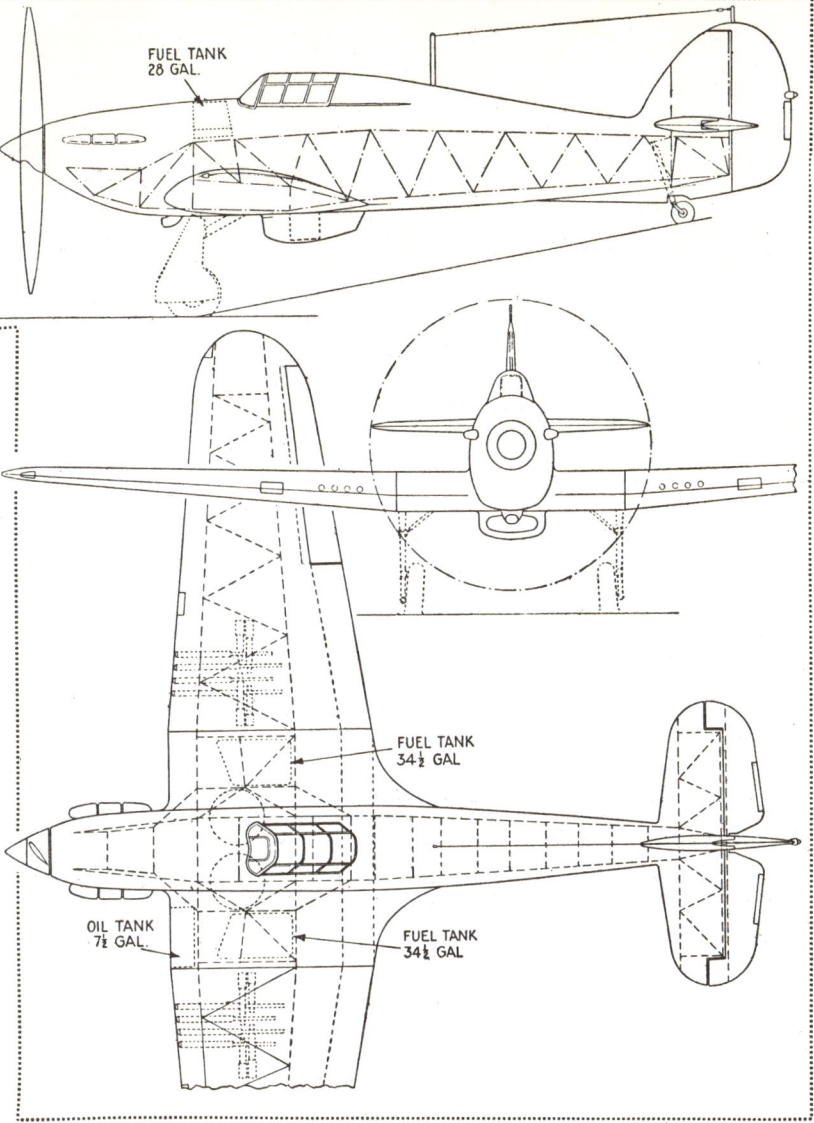

THE HAWKER HURRICANE
(990 1,050 Rolls-Royce Merlin Engine)

DIMENSIONS.

Length o.a.	31ft.	5in.
Wing span	40ft.	0in.
Span of tailplane	11ft.	0in.
Wheel track	7ft.	10in.

AREAS.

	sq. ft.
Main planes (gross)	257.5
,, ,, (net)	231.5
Ailerons (total)	19.6
Tailplane	19.78
Elevator	13.46
Fin	8.79
Rudder	11.86

FUEL TANK 28 GAL.

FUEL TANK 34½ GAL

OIL TANK 7½ GAL.

FUEL TANK 34½ GAL.

which is a single unit extending about three feet on each side of the fuselage, the web is solid and stiffened by riveted-on vertical channels. In the outer wing portions the web has circular lightening holes with their flanges turned out for stiffness.

It is, however, in the drag bracing that the Hurricane "motif" differs most from the biplane wings. The drag members are almost identical with the spars as regards their form and construction, but slightly smaller. They run zig-zag fashion between the spars, and are bolted to top and bottom spar booms via substantial forgings, which are fastened to the spar booms by horizontal bolts. Thus the primary structure of the wing forms a frame, braced and stiffened by the diagonal drag members. The resulting structure is enormously strong and, what is just as important, is remarkably stiff in torsion. As a manufacturing job this arrangement is very simple. The zig-zag drag members, held together by the forgings at their ends, form one unit. In vertical jigs, which are merely brackets on the wall, the two spars are attached and the primary structure is ready to receive the wing ribs.

These are of very simple type, and have ties or bracing members of circular-section tubes, while the flanges are simple channels with their free edges turned inwards. The bottom of the channel is not flat, but also forms a trough, the edges of which are rounded because the section is rolled or drawn from a single

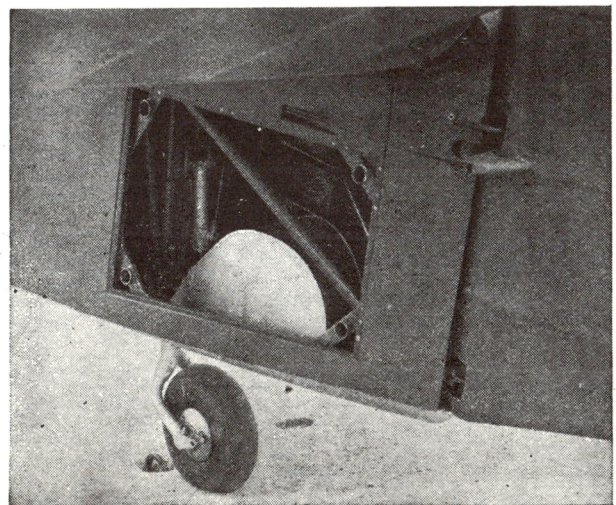

The tail wheel retracts into the casing seen in the stern of the fuselage.

strip. The result is that the outer edges do not cut the wing fabric, which rests on them. The attachment of the wing fabric to the ribs is extremely strong. When the fabric has been placed over the wing, a metal strip is placed over each rib channel and secured to it by Simmonds elastic stop nuts. As these are tightened up the fabric is drawn into the channel of the rib and instead of being held only at a few places, as in attachment by lacing, the fabric is held by the whole area of the strips.

Only over the inner wing portions where it is often necessary to walk on the wings is the covering of light-metal sheet. Here the wing rib channels have a plain smooth back for the attachment of the metal covering.

The nose and trailing-edge ribs are attached to the spar booms by horizontal bolts, and form separate units. Over the rear spar a short length of channel continues the contour of the rib, but on the front spar the ribs stop short at the spar boom. That is because the leading edge is covered with sheet metal, the rear edge of which passes over the spar booms.

Split trailing-edge flaps are fitted at the inner end of the wings. These are of the "single-surface" type, with a sheet-metal bottom surface stiffened by ribs from the operating torque tube. A notice in the cockpit tells the pilot not to use the flaps at speeds in excess of 120 m.p.h. The flaps are hydraulically operated. The ailerons are of the Frise type and are fabric covered.

The undercarriage is of the laterally retracting type. In the case of the Hurricane the wheels retract inwardly. By this arrangement a very wide wheel track is obtained,

When the large detachable panels are removed there is very free access to the Rolls-Royce Merlin engine. The large tungum pipe shown is part of the Glycol cooling system.

so that the machine is very stable on the ground and can be turned in a small circle at quite high speed without turning over. The telescopic leg, a Vickers oleo-pneumatic, is hinged to the bottom boom of the centre-section front spar. One member braces it in a fore-and-aft plane and another in a lateral plane. The hydraulic jack operates the latter, which is of the " broken " type. When this strut is " broken " by the jack, the telescopic leg swings inwards, and as the drag strut of the under-carriage is pivoted around a higher centre, it brings the wheel back as well as inwards. This is necessary in order to enable the wheel to clear the front spar when retracted. When the wheel is down it is in the plane of the spar. Fairings attached to the outside of each undercarriage close the opening in the wing surface when the undercarriage is retracted.

Two independent systems of operation are provided for the undercarriage. The main system is power operated and has hydraulic transmission. The auxiliary system is also hydraulic, but is hand operated. Should anything go wrong with both systems there is an arrangement of cocks whereby the pilot can release all pressure in the hydraulic systems and release the catches which hold the wheels up. The weight of the wheels then brings them into the " down " position.

The Rolls-Royce Merlin II engine is mounted on a simple steel tube structure in the nose of the fuse-lage. Large detachable panels in the cowling give access to most of the parts of the engine likely to need inspection or adjustment. The radiator of the liquid-cooling system is placed under the fuselage, with an oval entry in front of the radiator and a rectangular opening aft of it. The opening in the aft end of the radiator housing is covered with a hinged flap by means of which the pilot can vary the amount of cooling. An interesting feature of the cooling system is the use throughout of tungum pipes.

The split flaps of the Hurricane extend from radiator casing to root of aileron. They are hydraulically operated.

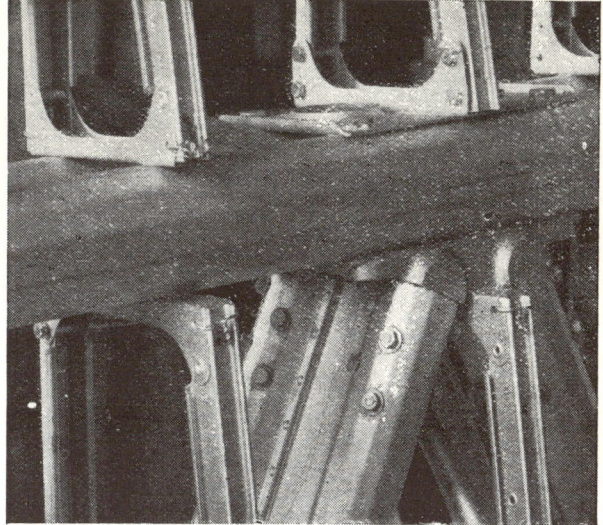

"Flight" photographs.
Forgings are used for attaching the diagonal drag members to the wing spars. On the left, a wing panel is seen in its jig.

Many have expressed surprise that the Hurricane is not fitted with variable-pitch airscrews. The answer is that it has such a tremendous reserve of power that even with the airscrew stalled at the beginning of the take-off run the acceleration is reasonably good, and the wooden air-screw is, of course, a great deal lighter.

Of the equipment and armament of the Hurricane little may be said. This machine is, as already mentioned, intended for day and night flying, so that it is provided with navigation lights, Harley landing lights and so forth. In addition to the latter it also carries two parachute flares. Complete blind-flying equipment is carried, and, of course, two-way radio.

The armament consists of eight Browning machine guns placed inside the wing. The idea is that in modern fighting the pilot will get but a very short opportunity for getting his target in his sights, and the multi-gun arrangement is designed to make the most of the short period. Firing of the guns is, of course, by remote control, which introduces a complication, but, on the other hand, there is no interrupter gear.

For the present, detailed performance figures cannot be given for the Hurricane. It should be realised that only the prototype has been fully tested at Martlesham ; certain minor modifications made in the production machines have resulted in a considerably improved performance. It has been authoritatively stated, as pointed out in last week's issue of *Flight,* that the Hurricane has the very remarkable speed range of 6 : 1. As the landing speed is about 60 m.p.h., and the actual minimum stalling speed is rather less than this, one is not far wrong in estimating the top speed at something like 330 m.p.h.

It is customary to give the duration of a fighter in terms of top-speed range. That is because the duration can obviously be increased very greatly by throttling. Probably the Hurricane carries petrol for nearly two hours at full throttle (it would not be of much practical use if it carried *much* less), and a top-speed range of about 600 miles appears a reasonable estimate.

A reassuring feature of the clean single-seater fighter is that it can remain in the air on very little throttle, so that when patrolling for invading bombers its duration could probably be increased to something like four hours. That is a point often overlooked when discussing the relatively short top-speed duration of fighters.

HURRICANES IN SERVICE

No. 111 (Fighter) Squadron shows off its new equipment

TRAVELLERS on Western Avenue, that famous (or notorious) London exit, are getting used to one of the most comforting manifestations of Expansion. No. 111 (F.) Squadron, at Northolt, is putting in all the time it can on its drab but deadly looking Hawker Hurricane multi-gun, "three-hundred-plus" monoplane fighters. Had it not been for Sqn. Ldr. Gillan's Turnhouse-Northolt trip at 408.75 m.p.h., the equipment of the squadron would most probably still be a State secret. Happily, it was decided that the world should be given some indication of the Hurricane's capabilities, though reports of strong aiding winds still kept the masses guessing at the full-throttle level speed of our first fighter to better 300 m.p.h.

There is no doubt whatever that No. 111 is, at the moment, the most formidable fighting squadron in the world. Although the actual armament of the Hurricane is secret, references have been made in the Press to the colossal weight of fire at the command of a Hurricane unit.

The engine is a Rolls-Royce Merlin II rated at 960/995 h.p. at 12,000ft.; maximum power is 1,050 h.p. at 16,000ft. at 3,000 r.p.m. Glycol cooling is employed, the system utilising a radiator in a scoop-shaped housing under the pilot's cockpit, and the airscrew is a fixed-pitch wooden two-blader. All three wheels of the undercarriage are retractable.

Full two-way radio equipment and night-flying gear is carried, and the pilot's cockpit is warmed and covered by

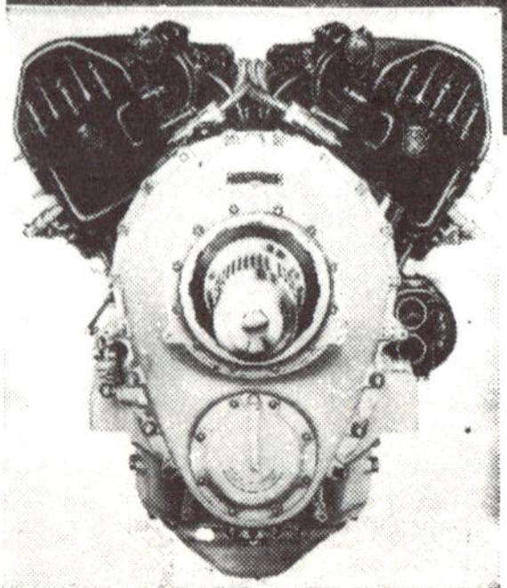

How the Rolls-Royce Merlin II permits clean entry and a generally compact installation in the Hurricane. The common housing for the oil and glycol radiators is apparent in the *Flight* photograph of the nose

12

a transparent sliding roof. The new service blind-flying panel is specified.

In view of the fact that there has lately been some heated talk about delay in the introduction of fighter monoplanes into the Service, and considering some of the fantastic reports of foreign equipment, *Flight* feels that it should be made clear that America has only a very small number of comparable (but poorly armed) Seversky machines actually in service, and that none of the new Curtiss monoplanes have been delivered; that France has very few, if any, Morane 405s with the squadrons; that Italy's single-seater fighters are mainly biplanes which compare unfavourably even with the Gauntlet, not to mention the Gladiator, and that, although Germany is known to have a number of Messerschmitts with Junkers engines on charge, the performance and armament of these machines probably show up rather poorly against the Hurricane. So we may be proud of No. 111 Squadron.

A fine study of perspective is offered by this "shot" of an echelon formation by Hurricanes of No. 111 (F) Squadron. The flying view opposite bears further testimony to the efficiency of the pilots

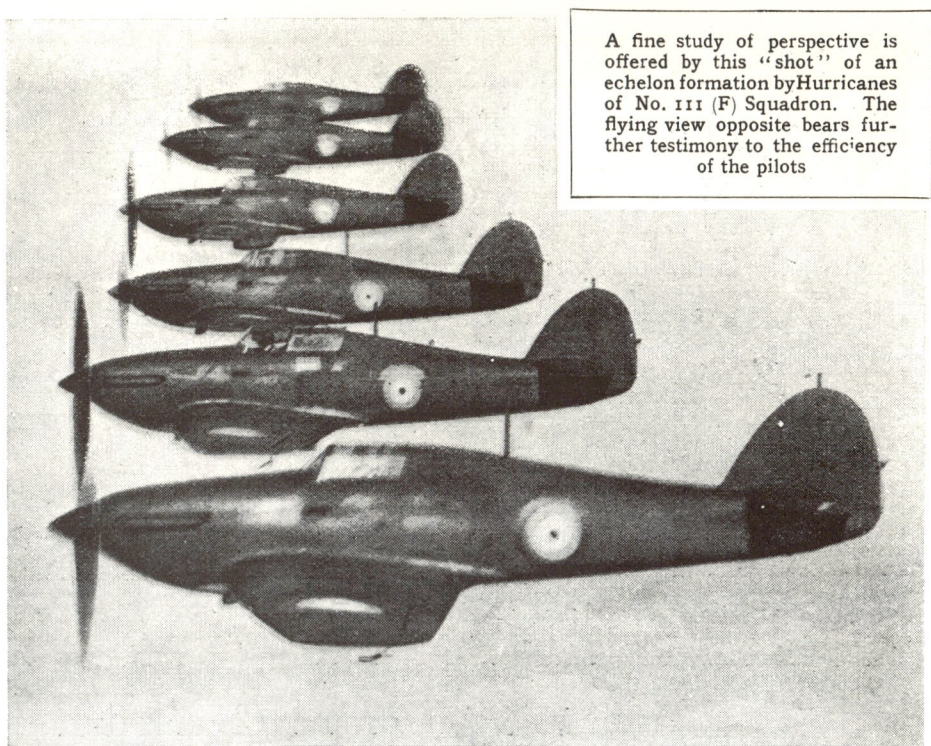

That the personnel is equal to the material was readily apparent at Northolt last Friday afternoon when the complete squadron demonstrated formation flying.

One was particularly impressed by the very reasonable landing speed of the Hurricanes, though the take-off *did* seem on the long side, particularly where formations were concerned. The wheels disappear without fuss, and once everything is snug the machine is fit and ready to challenge anything with wings.

A demonstration by an individual machine, though not aerobatic in nature, was a revelation of manœuvrability and awesome speed. One turn, particularly, must have generated a formidable amount of "g." It seemed to belie the story that the Hurricane makes a diameter of 3,000ft. in a loop.

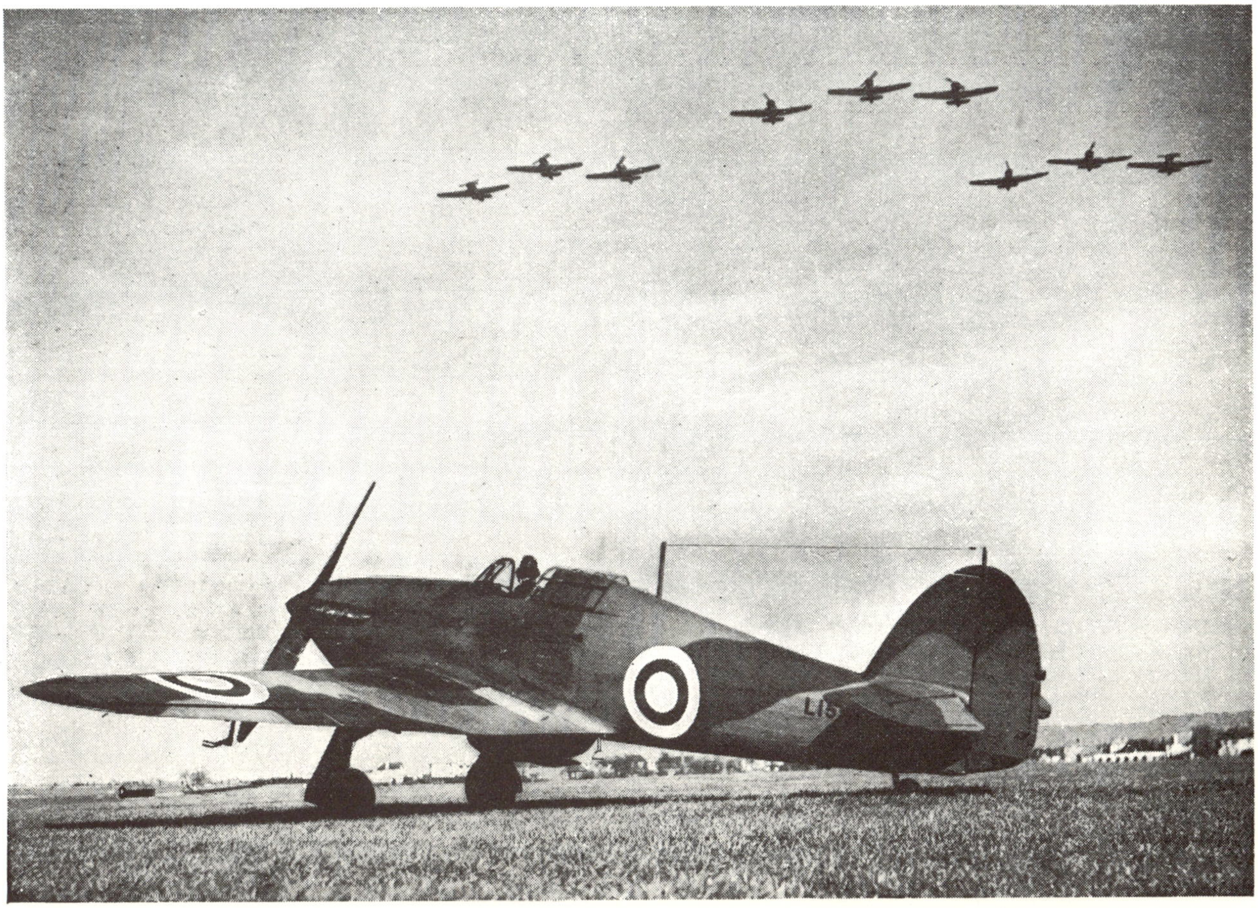

408 m.p.h. RECORD

by

R·A·F PLANE

"One of the new Hawker Hurricane fighter aeroplanes (Rolls-Royce Merlin Engine) of the Royal Air Force flew 327 miles from Turnhouse Aerodrome near Edinburgh, to Northolt, Middlesex, in 48 minutes—a speed of 408.75 m.p.h., or nearly seven miles a minute.

The flight was made under ordinary Service conditions, with full military equipment and a normal load of petrol."—*Daily Telegraph 11/2/38*

ROLLS-ROYCE

Aero Engines

for

SPEED & RELIABILITY

EDINBURGH-NORTHOLT at 409 m.p.h.

Sqn. Ldr. Gillan's Own Impressions

A GRAPHIC account of the flight last February from Turn-house (Edinburgh) to Northolt in 48 minutes ground to ground in a Hawker Hurricane is given in the *Journal of the Royal Air Force College* autumn issue, the author being the actual pilot, Sqn. Ldr. J. W. Gillan, C.O. of No. III (Fighter Squadron). We reproduce the following extract with full acknowledgments to our Cranwell contemporary. Sqn. Ldr. Gillan writes :—

"I was slumbering in the ante-room after lunch—a pleasant custom I had learned in the East ; it was a bitter day, a gale was blowing, and the clouds were at 500 feet. All our aeroplanes were in the hangars and there appeared no prospect of our being disturbed ; all our paper work was complete, and I had heard that the telephone system in the Camp had broken down. There were lots of coal on the fire and I remember wondering in my sleep whether it was not one of those days when one smoked one of the fast diminishing supply of cigars one had received as Christmas presents, when I was rudely awakened and shown a paragraph in a technical paper : 'Hurricane Flies from Edinburgh in Sixty-five Minutes.'

"The afternoon was spoiled. I remembered reluctantly the full-power trial which had to be made.

"I went to the telephone, called the Air Ministry to hear about the weather. They spoke of everything from ice formation to clouds covering the high ground ; but they did say wind was from the north at 75 miles per hour at 17,000 feet.

"This was the end of an afternoon of leisure.

"I went down to the hangars at 2 p.m. ; I had the aeroplane pulled out and started, and sent a telegram to Turnhouse asking for petrol. I was just going to leave when I was called back to the Mess—'wanted by the Air Ministry' ; the Press Section had noticed the announcement in the Technical Press. When was the full-power trial to be done ? This afternoon, they were told, and requested not to ask questions. I wanted to be back by dusk and I promised to ring them up on my return, and flew off to Edinburgh.

"I have read much about the return journey ; but nothing about the journey up, which was much more hazardous. As the wind was much stronger than I expected I arrived at Turn-house at 4.15 p.m. instead of 3.45 p.m. I found no petrol waiting, as I had arrived sooner than my telegram ; but owing to the efficiency of the Station this very necessary require-ment was forthcoming quickly. . . .

"It was now approaching dusk ; there were no clouds in Edinburgh and the sky was that dark blue that precedes the dusk. There was a northerly gale blowing and I decided that I would fly back, as I felt it would be im-possible to miss London in the dark. Taking off at 5.5 p.m., and climbing at 200 m.p.h., I noticed with some pleasure that I had no drift, and at 5,000 feet that I had a considerable sensation of speed. This meant a good wind and less chance of running out of petrol in the dark near London in the event of a miscalculation.

"The ground then disappeared and soon I was at my height with only my instruments, and a rapidly tiring left leg. My air speed varied between 305 and 325. The engine revolutions were constant at 2,975, and boost constant at 5½ lb. There were ten minutes of high cloud to go through, when the cabin frosted up, and a hoar-frost formed on the wings.

"Sometimes I felt sorry that I was doing this, and thought of the comfort of my men at Northolt ; at other times I felt glad. After forty minutes I decided to descend. The air speed now was 400 ; the revolutions 3,600 ; the ground speed was probably 550. I had an odd feeling flying through a cloud at night at a speed I knew to be in excess of 500 miles per hour. Coming out of a cloud at 5,000 feet, I saw momentarily a red light flashing the letter of my station. But by the time I had registered this I was seven miles farther on.

"The signal time from the take-off to going over Northolt was 43 minutes ; the time on my own watch 44 minutes.

"I returned to Northolt about five minutes later and landed."

A FAMOUS FOREBEAR.—Sopwith Camel single-seat fighters flying in formation over France in 1917. The Camel, a famous ancestor of the Hurricane, was one of the foremost fighters of the last War as the Hurricane is of this—23 years later.

HAWKER HURRICANE single-seat fighters of the Royal Air Force have seen more action in this War than has any other type and have shot down more enemy aeroplanes than any other fighter. In fact in a year of War Hurricane squadrons had scored something like 1,500 confirmed victories over German fighters and bombers. Hurricanes were the only British monoplane fighters to go to France, where they bore the brunt of the German offensive. Now in Great Britain Hawker Hurricanes are in the first line of defence.

The success of the Hurricane in a year of War has placed it among the great aeroplanes of all time—ranged in history with that great fighter of the last War, its famous forebear, the Sopwith Camel. For Hawker Aircraft Ltd. of to-day is the direct successor of the Sopwith Co. and Mr. T. O. M. Sopwith is still a Director of the Company.

Hawker Aircraft Ltd., through its fine designing team, came to the forefront among the World's builders of aeroplanes when it turned out the famous Hart series, more varied in purpose and adaptation than any line evolved from a single prototype. The long list of Hart derivatives—the Audax, the Osprey, the Hind, the Hardy, the Hart Dive Bomber, the Hart Trainers, the Demon and the Hector—is eloquent of its qualities. But the Hurricane has made its name on active service and so has come triumphantly through an ordeal with which no peacetime design can ever compete.

The Line of Descent

Thus, although the Hurricane broke new ground, it has years of successful single-seat experience behind it. That experience may account for much of the wise provision which has been built into its design, features capable of assuring its superiority over the German types which have not the same background. For the Hurricane can trace its ancestry back to the little Sopwith Tabloid single-seater which made such a sensation in 1913, the forerunner of all

the tractor " Scouts "—built for the great pilot Harry Hawker.

Then came the little Pup with the 80 h.p. Le Rhone, one of the most delightful aeroplanes to fly that was ever built—reputed to have been drawn out by Mr. Sopwith and Mr. Sigrist in an inspired moment on the floor of the design office at Kingston. It was succeeded by the Camel, the great fighter of 1917, tricky to fly but, once mastered, able to make circles around anything in its day. The Camel was followed by the Dolphin, a heavily armed single-seat fighter of high performance for its time, by the Salamander trench-strafer and then by the Sopwith Snipe, a beautiful little fighter which continued in service with the R.A.F. until 1925. The Snipe was the last great Sopwith fighter, for in 1920 the concern changed its name to the H. G. Hawker Engineering Co. Ltd. and in 1933 to Hawker Aircraft Ltd. The Woodcock, the Heron, the Danecock, the Hornbill, the Hawfinch, and then the Hawker Hornet carried on the fighter tradition.

The Hornet, which was rechristened the Fury when it was adopted by the R.A.F. in 1930, was one of the first aeroplanes to use the then new Rolls-Royce F engine, later called the Kestrel. The Fury was built in quantity and formed the equipment of, among others, No. 1 Fighter Squadron which this year has distinguished itself on Hurricanes with the Advanced Air Striking Force in France. A development of the

1913—The Sopwith Tabloid.

THE SOPWITH LINE.

1915—The Sopwith Pup.

1917—The Sopwith Camel.

1918—The Sopwith Snipe.

HURRICANE

FURY FORMATION.—Another famous Hawker single-seat fighter, the Fury, was the immediate predecessor of the Hurricane. The Fury may yet see active service for numbers of them are still flying at training establishments and in South Africa.

["Aeroplane" photograph]

Fury was the Nimrod, the Fleet Fighter version. In this machine the wings were enlarged and provision was made for catapulting and floatation gear. The Hawker Nimrod was for some time the standard single-seat fighter of the Fleet Air Arm.

The Evolution of the Hurricane

The design of the Hurricane was essentially the product of team work,—a co-operative effort by a loyal staff working under Mr. Sydney Camm, Hawker's Chief Designer. Among them should be specially mentioned Mr. R. H. Chaplin, Mr. S. D. Davies, Mr. R. McIntyre and Mr. R. Lickley, although to single out any one leaves many others unmentioned. The fine production efforts of the firm under the direction of Mr. R. W. Sutton should receive their due acknowledgment.

While they were developing the Fury the Hawker Design Staff was working on designs of a new monoplane fighter—the first monoplane which Hawker's had attempted.

In its early stages in the Spring of 1933 the design for the aeroplane which eventually became the Hurricane was known in the Hawker design office as the "Fury Monoplane." The early General Arrangement drawing, dated December, 1933, shows a graceful little low-wing monoplane with a fixed cantilever spatted undercarriage. It was designed for the 660 h.p. Rolls-Royce Goshawk steam-cooled motor, which accounts for

the absence of a radiator. In plan form the wing is remarkably like that of the Hurricane. The design loaded weight was only 3,807 lb.—the span 38 ft.

In these first stages the machine was thought out as a Private Venture design. The name Fury Monoplane was dropped and the name Hotspur was provisionally adopted. Afterwards the name Hotspur was used for another design—a two-seat fighter of which only the prototype was built.

In 1934 the design crystallised when Rolls-Royce Ltd. produced their P.V.12 engine which had an output of 1,025 h.p. at 15,000 ft. This motor later became the Merlin "C" and then the Merlin II, the engine installed in the Hurricane I.

About the middle of 1934 the Air Ministry specification F.36/34 was designed to fit in with the general layout of the new Hawker monoplane which had by then acquired a retractable undercarriage. And as the F.36/34 the Hurricane was known until it went into production early in 1936.

The decision to build the prototype was taken in December, 1934, and the first machine was flown for the first time, only eleven months later, on November 6, 1935. Its retractable undercarriage and enclosed cockpit were both new to British fighters at that time. The Hurricane has undergone little major modification since then except for the change from fabric covered to stressed skin metal wings.

Although when it first appeared in public even experienced pilots believed that it must be difficult to fly, still more experience has proved that the Hurricane is in fact more easy to handle than some of the single-seat fighters which went before. This was no lucky chance but the result of careful thought and design which devoted itself not only to securing aerodynamic excellence, but also to providing robust construction combined with a comparatively low wing loading which in the prototype was little different from that on the then existing single-seat fighter biplanes. The aeroplane was deliberately designed to

1935—The Hawker Hurricane prototype.

THE HAWKER LINE.

1925—The Hawker Hornbill.

1929—The Hawker Hornet.

1934—The Hawker Super Fury.

["Aeroplane" photograph]

BIPLANE PEAK.—*A squadron of Hawker Fury Mk II single-seat fighters (640 h.p. Rolls-Royce Kestrel VIs) of No. 25 (Fighter) Squadron, in echelon formation, taken in June, 1937.*

["Aeroplane" photograph]

MONOPLANE TRIUMPH.—*Two flights of Hawker Hurricanes (1,030 h.p. Rolls-Royce Merlin III and de Havilland constant-speed airscrews) diving in formation.*

operate from poor aerodromes and was built for easy maintenance in the field.

The wide-track undercarriage, designed to fold up inwards, and the medium-pressure tyres both helped towards affording the good take-off from indifferent aerodromes. The Hawker steel-tube construction had long since proved its fine qualities in maintenance and it was preserved when the designer moved forward from the biplane to the monoplane in its modern manifestations.

Really rapid take-off was made possible in the later versions by controllable-pitch airscrews; and thus, three high qualities —good take-off, high rate of climb and wonderful manoeuvrability—combined with great fire-power account for the extraordinary successes scored by Hurricane squadrons of the R.A.F. in France and over Great Britain.

The prototype was designed to fly at a loaded weight of 5,700 lb. In the first production machine this was increased to 5,850 lb. and the first production Hurricane actually flew at 6,000 lb. loaded weight. Since then the various modifications have increased the all-up weight to 6,600 lb., at which

the Hurricane is now flying, although for special purposes it can be raised to more than 7,000 lb. All this increase in weight has been possible without any alteration of the primary structure,—a good example of the Hawker practice of allowing a fair reserve of strength in the initial design. The increases in weight have raised the wing loading from 22.8 lb. per sq. ft. to 25.6 lb. per sq. ft., which is still a moderate loading. It shows the wisdom of the original policy of designing for a low wing loading so that the flying of the Hurricane would not be too big a jump in piloting technique from the biplane fighters which preceded it.

In the original design the armament chosen was four machine-guns housed in the fuselage and firing through the airscrew disc with interrupter gear. Not till much later was this changed for the present familiar armament of eight guns, four in each wing, firing outside the airscrew disc without the complication of interrupter gear.

The Prototype

The building of the prototype—K.5083—was begun in

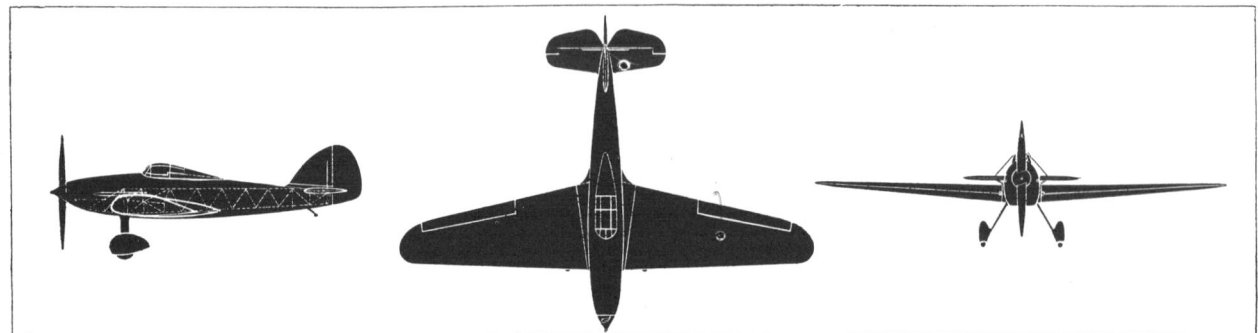

THE FIRST CONCEPTION.—*The layout of the P.V. Hawker Fury Monoplane of May, 1933. This aeroplane with a 660 h.p. Rolls-Royce Goshawk motor was never built but from it developed the Hurricane. The estimated speed of the Fury Monoplane was 280 m.p.h.*

[" Aeroplane" photographs

DEVELOPMENT IN DESIGN—VII.—How the Hawker Hurricane has developed from the prototype of November, 1935, down to the latest Hurricane Mk. I with the Rotol constant-speed airscrew. From top to bottom, the Hurricane prototype, the first production model, the modified production model with anti-spinning under-fin and fixed-pitch airscrew, and bottom, the latest standard Hurricane.

ON THE PRODUCTION LINE.—*A few of the Hawker Hurricane single-seat fighters on the assembly line at one of the Hawker Siddeley factories. The wide aisle between the lines prevents congestion and so speeds production. Although the spaces between fuselages appear small, they afford room enough for movement and for all the work associated with assembly at this stage.*

October, 1934. The first flight in it was made by Flight Lieutenant P. W. S. Bulman at Brooklands on November 6, 1935. At that time the Hawker Monoplane Fighter, as it was called, would have been almost unrecognisable to eyes now accustomed to drab camouflage. Then it was resplendent in silver dope and polished cowling.

From the first the machine flew well and on December 4, 1935, it was first mentioned in public in THE AEROPLANE. Flight Lieutenant Bulman felt so confident of its handling qualities after only a few hours of tests that he flew in close formation on the Hawker Hart to afford close-up photographs. As the first British single-seat monoplane fighter with a fully retractable undercarriage it created a sensation. That retractable undercarriage caused some qualms before the machine flew as it was the first which Hawkers had attempted. In fact, there has never been any trouble with it. This may be partly because the undercarriage was designed so that it could be extended entirely by gravity should the hydraulics fail.

The photographs taken in December, 1935, show tail bracing struts which were fitted at the start as a precaution against tail buffeting. They were soon removed as unnecessary. At this time the tail wheel was retractable.

Much pioneer research work was done with the prototype K.5083 as there was hardly any existing data on high-speed military monoplanes at that time. The speed of the prototype with the 1,025 h.p. Merlin C was 325 m.p.h. at 16,500 ft.

The Hurricane was put into production early in 1936 and the first production machine L.1547 flew in October, 1937. It differed but little from the prototype. The biggest modification was the installation of the 1,030 h.p. Rolls-Royce Merlin II in place of the Merlin C. The new motor compelled some changes in the cowling lines, the engine mounting and the cooling system. The radiator duct was redesigned with the characteristic " lip " which lowered the entry opening by two inches to avoid picking up the " tired air " in the boundary layer. The position of the radiator under the centre-section of the wing was selected to take the maximum advantage of the slipstream effect.

CONTINUED ON PAGE 26

HAWKER HURRICANE

Scale
0   5   10 feet

IDENTITY IN OUTLINE.—*The lines of the standard Hurricane Mk. I (1,030 h.p. Rolls-Royce Merlin III, Rotol constant-speed airscrew and ejector exhausts). These drawings form an interesting comparison with those of its original conception in design, the Fury Monoplane or Hotspur, on page 256.*

Characteristics of THE HAWKER HURRICANE Mk. I Single-Seat Fighter
(1,030 h.p. Rolls-Royce Merlin III motor).

DIMENSIONS

Span .. .. 40 ft. 0 ins.	Track .. .. .. 7 ft. 10 ins.
Length .. .. 31 ft. 0 ins.	Airscrew Diameter .. 11 ft. 0 ins.
Height, one blade vertical, tail down, 13 ft. 1½ ins.	Aspect Ratio 6.2

AREAS

Wing (gross) .. 257.5 sq. ft.	Ailerons (total) .. 19.6 sq. ft.	Tailplane .. 19.75 sq. ft.
Wing (net) .. 231.5 sq. ft.	Flaps .. .. 24.0 sq. ft.	Elevators .. 13.5 sq. ft.

WEIGHTS

Empty (bare) .. .. 4,911 lb.	Pilot and parachute .. .. .. 200 lb.
Fixed equipment .. .. 316 lb.	Ammunition .. .. .. 159 lb.
Movable equipment .. 357 lb.	Fuel .. .. 589 lb.
	Oil .. .. 68 lb.
Empty (equipped) .. .. 5,584 lb.	Disposable load .. .. .. 1,016 lb.
Loaded weight (gross) .. .. 6,600 lb.	

LOADINGS (At 6,600 lb.).

Wing .. 25.6 lb./sq. ft.   Power .. 6.4 lb./h.p.   Span .. 4.12 lb./sq. ft.

PERFORMANCE

SPEEDS—Maximum .. 335 m.p.h. at 18,500 ft.	At ground level .. .. 272 m.p.h.
*RANGE—At 5,000 ft. .. 830 miles at 168 m.p.h.	At 20,000 ft. .. 730 miles at 213 m.p.h.
At 10,000 ft. .. 800 miles at 180 m.p.h.	At 25,000 ft. .. 695 miles at 232 m.p.h.
At 15,000 ft. .. 775 miles at 196 m.p.h.	

CLIMB—Time to 5,000 ft. = 2.1 mins.   15,000 ft. = 6.5 mins.   25,000 ft. = 13.1 mins.
         10,000 ft. = 4.3 mins.   20,000 ft. = 9.3 mins.

Max. rate of climb = 2,420 ft./min. at 11,000 ft.

Service ceiling 35,000 ft.   Absolute ceiling 36,000 ft.

*The ranges tabulated above are based on the fuel remaining after allowance has been made for the fuel required for run-up, take-off and climb to operational height.

NASAL SURGERY.—How the development of the Hawker Hurricane has affected the lines of the nose. The top row shows the versions with two-blade fixed pitch airscrews, the second row the three-blade c-p successors. Top row, left to right: The prototype Hurricane, with stub exhausts, early type wheel fairings and original cockpit enclosure, Nov., 1935; the first production Hurricane, with flame-trap exhausts, new wheel fairings and cockpit cover, Oct., 1937; modified production model, with ejector exhausts, June, 1938. Bottom row, left to right: the first Hurricane with a controllable pitch airscrew, a de Havilland two-pitch type, Oct., 1938; the first Canadian-built Hurricane with a de Havilland constant-speed three-blade metal airscrew, Jan., 1940; the latest Hawker Hurricane Mk. I with Rotol constant-speed three-blade wooden airscrew, June, 1940.

Built-in Mass-Balance

Fabric covered Rudder

Rudder Post

Metal Leading-Edge

Fabric Covering

Tail-Light

Reinforced Section (Anti-Crash)

Wooden Formers & Stringers

Upward Identification Lamp

Balanced Trim Tab

Trim Tab

Armour Plating

Fixed Trim Tab

Two-Spars & Diagonal Bracing

Metal Leading-Edge

CLARK

'Dowty' Fixed Tailwheel

Handling-Bar Socket

Wooden Underfairing

Tube Framing (Wire Braced)

DETAIL OF COOLING SYSTEM

Glycol Header Tank
2 gal. Coolant
2 gal. Air Space

From Motor Cyl Blocks

Temp. Gauge

Relief Valve outlet through Cowling

Thermostat & By-pass

Inlet to Motor

Oil Cooler Inlet & Outlet

Air Flow Control Shutter

Glycol

Oil

AIR FLOW

Glycol

Flap Rod Universal Joint

Metal Flaps

Back Spar

Spar changes section

Fabric-covered Ailerons

Rib Formers

Back Intermediate Spar

Spar changes section

Landing Light

16 gauge reinforc plating (Torsion-Box Structure)

Stringers

Stressed-skin (metal) covering

Front Main Spar

Forward Intermediate Spar

Detachable Wing-Tip

Wingtip Nav. Light

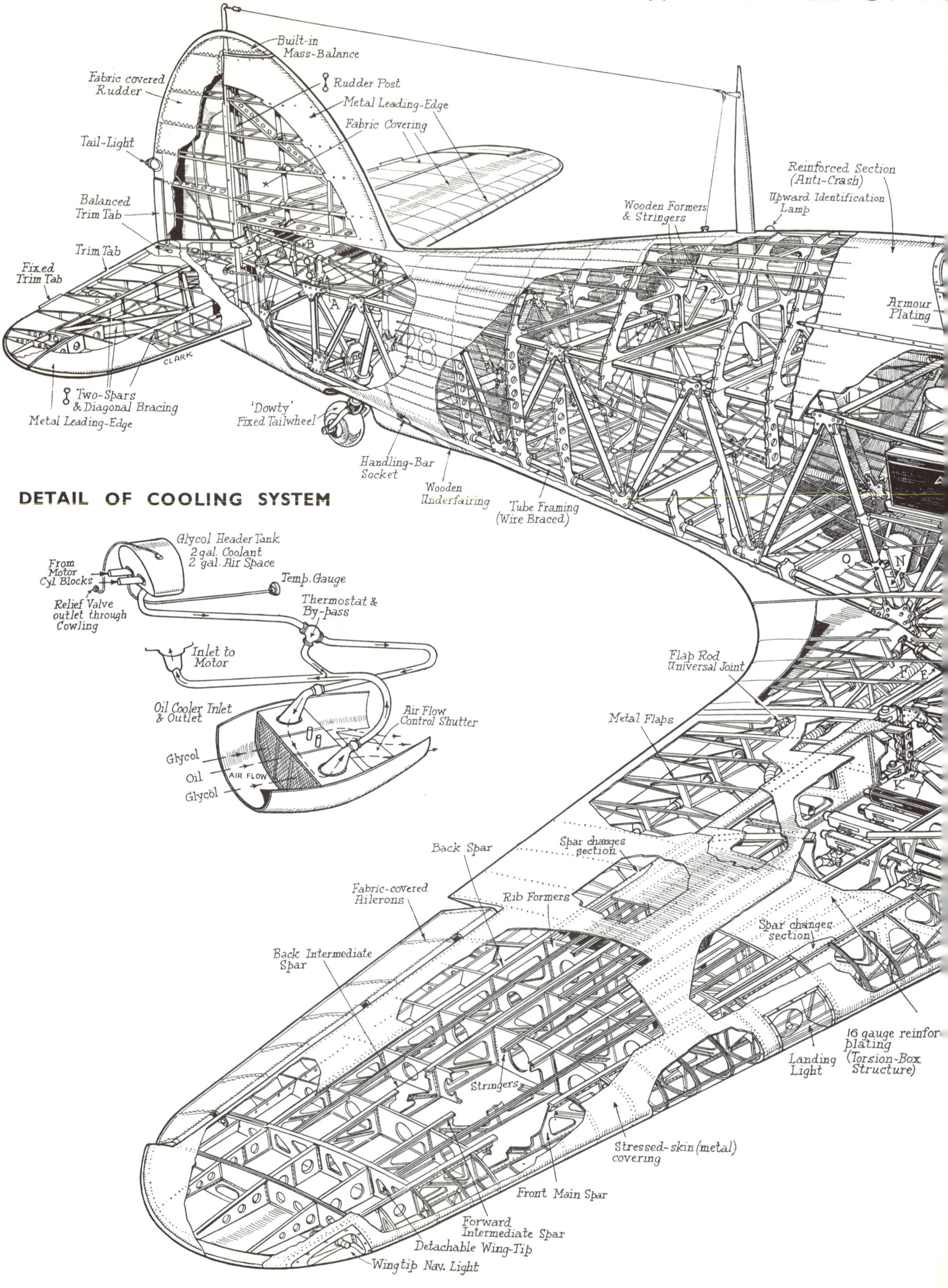

(1,030 h.p. Rolls-Royce Merlin III motor and Rotol three-blade Constant-speed wooden airscrew)

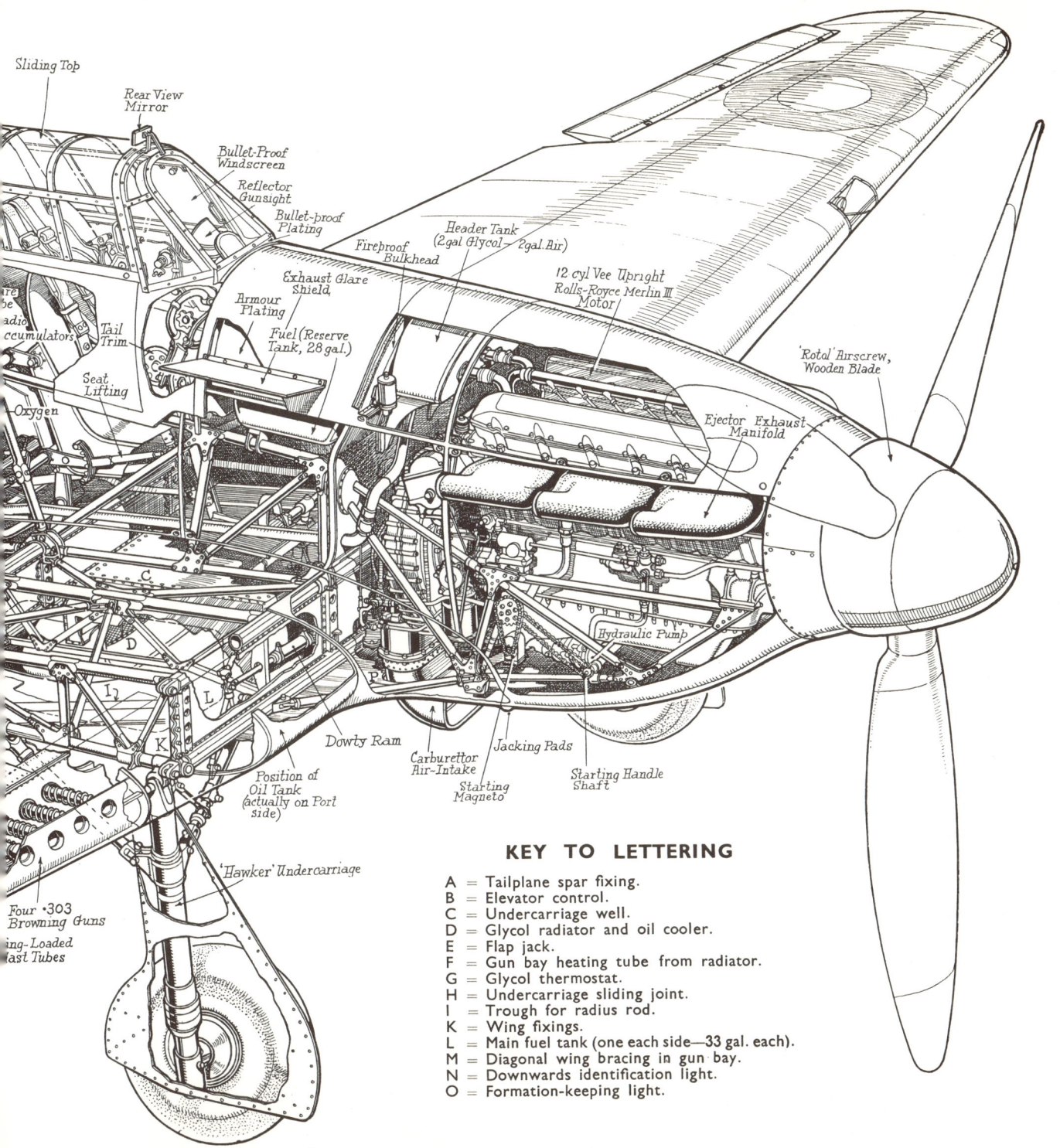

Sliding Top

Rear View Mirror

Bullet-Proof Windscreen

Reflector Gunsight

Bullet-proof Plating

Header Tank (2 gal. Glycol — 2 gal. Air)

Fireproof Bulkhead

12 cyl. Vee Upright Rolls-Royce Merlin III Motor

Exhaust Glare Shield

Armour Plating

Fuel (Reserve Tank, 28 gal.)

'Rotol' Airscrew, Wooden Blade

Tail Trim

Seat Lifting

Oxygen

radio

accumulators

Ejector Exhaust Manifold

Hydraulic Pump

C

D

I

L

K

P

Dowty Ram

Carburettor Air-Intake

Jacking Pads

Starting Magneto

Starting Handle Shaft

Position of Oil Tank (actually on Port side)

'Hawker' Undercarriage

Four ·303 Browning Guns

ing-Loaded last Tubes

KEY TO LETTERING

A = Tailplane spar fixing.
B = Elevator control.
C = Undercarriage well.
D = Glycol radiator and oil cooler.
E = Flap jack.
F = Gun bay heating tube from radiator.
G = Glycol thermostat.
H = Undercarriage sliding joint.
I = Trough for radius rod.
K = Wing fixings.
L = Main fuel tank (one each side—33 gal. each).
M = Diagonal wing bracing in gun bay.
N = Downwards identification light.
O = Formation-keeping light.

'The Aeroplane" copyright drawing.

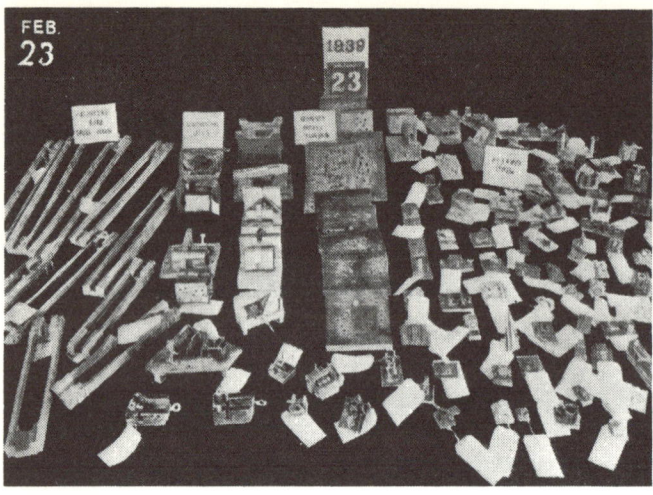

FEBRUARY 23, 1939.—Jigs prepared or the construction of the first Canadian Hawker Hurricane are assembled at the Montreal plant of the Canadian Car and Foundry Co. Fuselage tube drill jigs are on the left, welding jigs on the right and gusset drill plates in the middle.

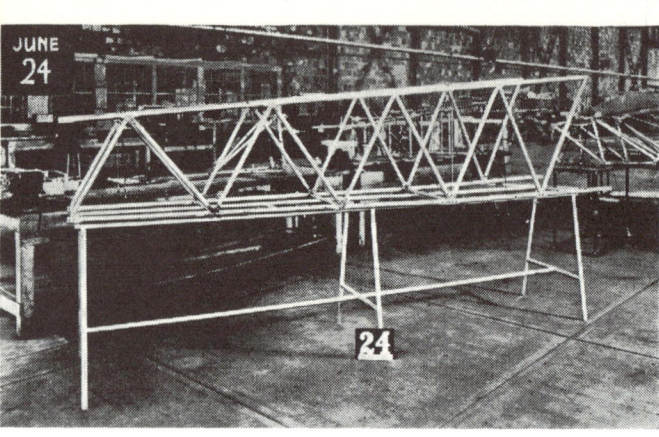

JUNE 24, 1939.—The aft part of the Hawker steel tube fuselage in process of assembly. The patented Hawker system in which the ends of the tubes are made square for ease in connection is an essential feature in speeding production.

THE BUILDING OF THE

OCTOBER 30, 1939.—Another month and the Rolls-Royce Merlin is set in the engine bearers and the undercarriage is in place. The radiator is fitted and the main fuel tanks are being installed in the wings. Much of the "plumbing" has already been done and a start has been made on the wiring. The tail unit and tail wheel are fitted.

NOVEMBER 21, 1939.—A great deal of detail work is done at this stage. The cockpit cover is completed, the reserve fuel tank fitted and all the complicated cockpit installations put in. The oil tank can be seen in the leading edge of the stub wing. The final connections of drives and control wires have mostly been made.

CANADIAN CAR AND FOUNDRY CO. of Montreal, Canada, has a contract with the Canadian Government for the building of Hawker Hurricane Mk. I single-seat fighters (1,030 h.p. Rolls-Royce Merlin II motors). Production is well under way. Fighter Squadrons of the Royal Canadian Air Force are now on active service in this country flying Canadian-built Hurricanes and have scored many confirmed victories. These photographs show stages in the construction of the first Hawker Hurricane to be built in Canada. Work began on February 23, 1939 and the first Canadian-built Hurricane made its initial test flight on January 10, 1940—less than eleven months after work began. 660 are on order.

JANUARY 9, 1940.—Now completed, P.5170 is wheeled out on to the aerodrome for engine running tests and final checks on brakes, flaps and controls before the flying trials.

JULY 29, 1939.—*The steel tube framework of the fuselage is practically complete from stern post to engine-bearers, forming the backbone of the aeroplane. Meanwhile the assembly of the wings is proceeding.*

SEPTEMBER 30, 1939.—*Seven months from the start the Hurricane is beginning to take form. The centre-section of the wings has been offered up to the fuselage, which now has its fairing stringers in place around the main structure of the fuselage.*

HURRICANE IN CANADA

DECEMBER 19, 1939.—*The Hurricane approaching completion. The fuselage and stub wings have been covered, whilst elsewhere the main-planes are practically complete. The undercarriage fairings and hydraulic retracting mechanism have been added. The wings are attached experimentally and then removed for transport.*

JANUARY 8, 1940.—*The completed fuselage of the Hurricane is towed on its own wheels from the Montreal plant to the St. Hubert airport for its trials. The de Havilland three-blade controllable-pitch airscrew has been fitted to the Merlin. In the airport hangar the wings are attached.*

JANUARY 10, 1940.—*The first Hawker Hurricane to be built in Canada takes off on its first test flight—the herald of a steady stream of similar fighters for the defence of the Mother Country.*

HAWKER HURRICANE single-seat fighters are designed not only for speed, manœuvrability, fire-power and ease of handling, but also for quickness and simplicity in production. The well-tried Hawker system of steel tube construction, combined with well-thought-out detail design, accounts for the speed with which Hurricanes are being turned out in Canada as in Great Britain. The eleven months taken by the Canadian Car and Foundry Co. Ltd. to build its first Hurricane in Canada was, in fact, exactly equal to the time which Hawker Aircraft Ltd. took to build the prototype Hurricane in 1934-35. Since then, the rate of output has risen steadily in both countries.

CONTINUED FROM PAGE 20

At the same time as these modifications some detail improvements were built into the design. A more rounded windscreen and cabin top were fitted. Streamlined exhaust manifolds instead of stub exhausts were installed. The mass balance for the rudder was enclosed instead of being external as on the prototype. The hinged portions of the leg fairings of the undercarriage were removed for simplicity's sake. This made practically no difference to the performance, but gave the Hurricane its characteristic look of having tucked up its toes when seen from underneath.

Eight Browning machine-guns and night-flying and navigational equipment were installed. The guns were grouped together compactly, four on each side outboard of the airscrew disc. In fact the wings were designed round the gun bay to gain the maximum efficiency.

After the Hurricane had been in service for a short while the tail wheel was made non-retractable for further simplicity. In this form with a two-blade fixed-pitch wooden airscrew the top speed of the Hurricane was 330 m.p.h. at 17,000 ft.

Since it first went into squadron service with No. 111 (Fighter) Squadron, R.A.F., at Northolt early in 1938 modifications have been made which distinguish the latest Hurricanes from the earlier models. First of all there was the fitting of the anti-spinning fin under the fuselage. Then came the introduction of the ejector-type exhaust manifolds which gave an increase in the top speed of about 10 m.p.h. and a corresponding gain in cruising speed and faster climb. More recently the fitting of the de Havilland and Rotol three-blade constant-speed airscrews has greatly improved the take-off and climb and added a further 5 m.p.h. to the top speed.

Most important of all from a structural point of view the metal-framed fabric-covered wing has been replaced by an all-metal stressed-skin wing of great structural efficiency. The new pair of wings is about 70 lb. lighter than the fabric-covered wings and has much greater stiffness although the section and plan forms remain unchanged.

In this latest form, with metal wing, ejector exhausts and Rotol constant-speed airscrew the Hurricane I has a top speed of 335 m.p.h. at 17,500 ft.

The Structure

The structure of the Hurricane is essentially straightforward and designed for simplicity and efficiency both in production and in service.

The wings are divided into three main components, the centre section and the two outer wings. The centre section consists primarily of two continuous spars, the polygonal rolled-steel booms and steel plate webs, connected by ribs and drag bracing. It houses the retractable undercarriage and main fuel tanks.

The outer wings of the prototype and early production Hurricanes were built up on two main spars similar in construction to the centre-section spars, and braced torsionally by a system of strong diagonal members between the spars. The wings were fabric covered over a framework of light metal ribs. This wing construction was described in THE AEROPLANE of May 11, 1938. In 1938 a completely new stressed-skin outer wing was designed. It was introduced on the production Hurricanes in 1939.

The new outer wings are built up on two main spars of extruded π-section booms with double webs at the inboard end which are reduced to T-sections with single webs towards the tip. A diagonal bracing system in the gun bay was found to be convenient. The diagonals are now made of T-section extrusions with plate webs. Immediately outboard of the gun bay the wing is of the fully-stressed-skin type and has two light auxiliary spars as well as the two main spars. The whole wing is covered with a stressed skin of light alloy, 20G. thick on the top surface and 22G on the bottom. The ribs are pressed from sheet.

The loads are diffused from the diagonally braced gun bay to the stressed-skin portion through a torsionally stiff box rib made up of two heavy gauge plate ribs with extruded T-section booms immediately outboard of the gun bay. There is an extra covering of 16G light alloy as a reinforcement underneath the normal skin covering.

The ailerons, which have a metal framework and fabric-covering, are built up on a tubular steel spar. They are the same as on the earlier wings.

The fuselage is built up on a basic rectangular structure of steel and light alloy tubes with squared ends. The tubes are connected by the standard Hawker method of flat plate fittings secured by tubular rivets or bolts. The whole is rigidly braced. This basic structure is faired to an oval section covered with detachable metal panels forward, and aft with fabric over light wooden formers and stringers.

The whole of the cantilever tail unit has a metal frame and fabric covering. The fin is built integral with the rear fuselage. The rudder is mass balanced. The tailplane is of the fixed type. The aerodynamically balanced elevators have trimming tabs.

The retractable undercarriage has two semi-cantilever Vickers shock-absorber struts hinged at the outboard ends of the centre section front spar, and retracted inwards by a Hawker mechanism actuated by Dowty hydraulic rams, which brings the wheels up between the centre section spars in the retracted position. A slight backward motion is provided by a hinged rear strut which slides on a guide at right angles to the spar of the wing. Emergency extension is by gravity. Dunlop wheels and pneumatic brakes are fitted as standard. The Dowty tail wheel unit is not retractable.

A FINAL GROOMING.—A new Hurricane from the shops is parked on the apron ready for the running trials of its Merlin motor before its first flight.

Power Plant

The standard 1,030 h.p. Rolls-Royce Merlin III 12-cylinder upright-Vee glycol-cooled motor is mounted on a tubular structure similar in construction to the main fuselage. The motor drives a Rotol three-blade constant-speed airscrew of 11 ft. diameter. The Rotol is now standard on the Hurricane, the D.H. airscrew on the Spitfire. Some Hurricanes have de Havilland airscrews and can be recognised by the more pointed spinner as compared with the rounded spinner of the Rotol. These Rotol airscrews have wooden blades as standard.

The fuel is carried in two main tanks in the centre section between the spars with a total capacity of 69 gallons. A further 28 gallons is carried in the reserve tank in the fuselage in front of the pilot. The oil tank, with an effective capacity of 7½ gallons, is in the port leading edge of the centre section. The tanks are protected by armour plate and rubber.

The main glycol radiator is housed in a duct under the fuselage below the cockpit. The oil cooler is sandwiched between the two elements of the glycol radiator.

The enclosed cockpit,—immediately over the centre section to give the pilot a good forward view,—has a sliding canopy. An emergency escape panel is built into the side of the fuselage between the upper longerons and the sliding hood. A bullet-proof windscreen and armour protection for the pilot are provided. The seat is adjustable for height, and the rudder bar for length of leg.

The eight Browning machine-guns mounted in the wings fire at a total rate of 9,600 rounds per minute. Complete night-flying equipment, with landing lights in the wing leading edge, navigation lights, oxygen, radio, flare tube, etc., is fitted as standard equipment.

With full tanks the Hurricane has a range of 830 miles at 168 m.p.h. cruising at 5,000 ft., conditions which would make it suitable as a bomber-escort on shorter flights. Cruising at 232 m.p.h. at 25,000 ft. the range is still 695 miles.

Having lived with the Hurricane for so many years the Hawker design staff is very modest about its performance and is eager to point out the ways in which it might be improved. We can expect that with all the experience gained in the development of the Hurricane the Hawker Co. will follow up its great success with another fighter of still better performance, capable of carrying on a reputation which ranks high among the great single-seat fighters of the World.

A Hurricane in the Hand

KING ARTHUR had his Excalibur, Hereward his Brain-Biter; the Baron of Shurland called his sword Tickletoby and Gymnast addressed his in terms which only Rabelais would have put on paper. These famous blades will have a rival in history, and few names could be more appropriate to the weapon with which our young men are scourging the Hun than the one it was given in some inspired moment: the Hurricane.

It is as beautifully balanced to the hand as Martin Lightfoot's battle-axe; and has a similar devil in the "handle" which makes one eager and able to kill. Like the sword which Weland made for Sir Richard Dalyngridge, it croons to its owner, but with the far mightier voice of a thousand Mare Swallows. It is steed and lance in one.

Apart from the inestimable rôle which the Hurricane has played and is playing in the history of Mankind—no less—it is also a mere aeroplane; and as one of these mechanisms we must consider it.

It is a pleasure to approach a Hurricane to fly it. It has that stark practicality and soundness in every line which one associates with the name of Hawker. The finish is most pleasing, the more so as they come out like sausages nowadays.

One rolls out with the seat notched up to its highest, so as to see as much as possible over the smooth, swelling expanse of brown cowl over the Merlin, which mutters at one through the three great spouts on each side. There must not be too much waiting about, for the radiator is designed to cool in a 200 m.p.h. slipstream or more, and not in the lazy draught from an idling screw. On the boundary the seat is let down, the trimmer set, the mixture confirmed in RICH, the throttle clamp tightened, the pitch set to FINE, the flaps tested and left closed and the radiator shutter closed half way.

The take-off is a great moment for those who like that long-drawn-out push in the back, the eagerness to be away of a thoroughbred fighter. For the first few yards one watches in case of a swing, for although the Hurricane is not prone to it this may happen to anything powerful.

Once unstuck the undercarriage must come up at once, and this entails taking the left hand from the throttle to the stick, and the right hand from the stick to the selector lever. This is the only awkward feature in the handling, and as the elevators are light and effective at low speeds the new man on a Hurricane can usually be noticed plunging a bit.

In the old type the hydraulic lever then had to be held down while the wheels came up and locked home, so a wise preliminary to the changing of hands was to throttle back slightly to avoid over-revving in fine pitch. To help in this a steep climb was usually held till the wheels were up and the hand free to change pitch. Nowadays the undercarriage goes through the motions without other action than selecting "Wheels Up," and the airscrew is of the constant-speed persuasion so can be set to the best r.p.m. at once.

About this time the pilot usually closes the hood, taking good care when reaching back that his elbow does not protrude into the slipstream, or he may wrench a shoulder badly. The goggles may then be put up and the seat raised, whereupon the view becomes admirable. The cockpit is completely comfortable apart from a tendency to over-warmth on a Summer day low down.

The secret of the Hurricane's immense effectiveness cannot be fully analysed, but much of it must come from its extraordinary handleability. The wing-loading is moderate and the buoyancy immense. At ordinary cruising, which is naturally well over 200 m.p.h., all controls are just heavy enough not to be "tricky" yet so exhilaratingly lively in response that one is tempted into tight manœuvres up to blacking-out point. One ferry pilot of sober years returned after his first Hurricane and reported that he was still in a beautiful dream and practically jet black.

A touch back on the stick and every pointer on the instrument board which means "Up" goes round and round; yet in straight flying the Hurricane is stable and gentle beyond most touring craft.

Some miles before arriving one throttles back, then the speed may have dropped to 150 m.p.h. in time to lower the wheels before circling. When the wheels are down and the green lights show the flaps are also lowered partly and the hood slid back. All is then in order for a slow circuit at about 100 m.p.h.

The final approach is done at 80 m.p.h. with full flap and airscrew in fine pitch. Although this is a perfectly kind and manœuvrable aeroplane at that speed, and the stall only arrives at considerably less, the correct practice is a trickle of engine to govern angle of descent. The landing itself is a sit and a roll which seems almost to happen by itself.

"And pat he comes, like the vice i' the old play." Lord Haw Haw of Hamburg announces solemnly that our pilots are refusing to fly the Hurricane because it is vicious! Poor Haw Haw! His news gets richer and richer in half-digested leg-pulls. But I suspect that German airmen have told him most earnestly and fervently that this lovely thing is a death-trap. So it is. And HOW! But not to *our* pilots. Ask them.—F.D.B.

NICE FLYING ANYWAY : The Hurricane has all the good flying qualities expected of a Hawker machine.

THIS IS THE HURRICANE

"*Indicator*" Records His Impressions of Flying a Nice Aeroplane

DESPITE the fact that the Hurricane was designed and originally developed so very long ago, and that there are now quite a number of new and altogether more savagely effective fighters on their way into service, this type will probably retain its special place in the scheme of things for quite a long time to come. Change the word "despite" to "because of" and you have the greater part of the reason why it has been, is, and will continue to be, such a success.

Time is required to bring perfection and to introduce gradually all those improvements which real work, both in peace and in war, prove to be necessary, while the new fighters are following slightly different trends. In short, the Hurricane and the Spitfire, each in its own particular way, are perfect for the work which they now have to carry out and against the types which are at present being used in quantity by the *Luftwaffe*. And they will continue to be perfect for this work while other fighters come on to the scene to perform slightly different functions and to fight rather different battles.

Since it was the first of its type, the Hurricane will always be remembered with special affection by those within the aircraft industry, on its fringes, and in the R.A.F., even when the war has produced other and much more remarkable aeroplanes. It seems to be a very long time ago that this machine first came into the limelight with a somewhat remarkable flight from Turnhouse to Northolt. Gale or no gale, it was a tremendous performance to average more than 400 m.p.h., and a still more tremendous performance to average this speed at night when the general public still looked on modern fighters as machines which should only be flown in the best possible conditions and had hardly realised that night flying was done by machines other than those used for long-range bombing and air transport, these necessarily carrying a suitable crew. The thought of one man alone behind a thousand horse-power and travelling at 400 m.p.h. at night gave us some new ideas.

In those days, too, the ordinary pilot outside the Service did not realise that such a high-speed night cross-country was normally possible largely because of the innate excellence of the Hurricane as a machine. It was not until more or less ordinary mortals were poured into the cockpits of these fighters and set off casually for delivery trips that the superman-test pilot idea began to fade. It has reappeared again, of course, with new and faster types, and will probably die in the same way. Once upon a time the Hart was considered to be the last word in military machines, to be flown only by the magnificently fit and well trained, and it is now treated as a sort of joke to be handled by pilots who are not yet fit to deal with anything more splendid.

Yet there is no doubt that, once the control drill of the modern type has been mastered, the Hurricane is a great deal easier to fly than the Hart or Hind. The only difference is that it is rather more lethal because of its higher wing loading and loses speed rather more quickly during an approach and landing because of that comparatively recent invention, the flap, without which no modern type could be flown. Incidentally, few modern types could be flown without some form of controllable pitch airscrew, and it says much for the design of the Hurricane that it was produced before such airscrews were reasonably available and was thoroughly effective in service without such an aid to take-off and climb.

A Private Theory

A nice compromise in wing loading and airscrew pitch was arranged, and this compromise has helped to make the Hurricane a fighter of all-round merit. Without the necessity for such consideration it might have been just a little on the "hot" side, and not, therefore, so suitable for handling by pilots with very little previous experience of high loadings—nor so suitable, either, for work in rough-and-ready conditions. One of my private theories about the development of air war is that whatever the advances in design, there will always be uses for improved versions of obsolescent types. Sooner or later it will be necessary to fight from a terrain where aerodromes are poor or non-existent, or where all the aerodromes have been effectively but not completely destroyed on both sides of the line. Then the advanced types will be grounded and the older

tions, the acceleration was the result of a too careless use of plus six boost, and the rudder corrections were quite unnecessarily vigorous.

At that time I was not by any means used to going quickly, and by the time I had lowered my seat, closed the hood, completed the undercarriage retraction, reduced boost and changed to coarse pitch—Hurricanes had two-position airscrews in those days—I was dashing along over entirely unfamiliar country at 200-odd and on a compass course which I hoped was correct. Of course, I hit a familiar landmark in due time (one always does) but it seemed to me then that it was a bit early in my flying career with military types to be forced landing for no good reason. After that I learnt, at least in a strange machine, to get everything set and fair before leaving the aerodrome. A Hurricane, in particular, turns by itself, and there is no hardship in making a circuit or two. Perhaps I wanted to impress everyone with my enthusiasm and confidence. I've forgotten.

Not that these modern types are

FEET PER SEC. : A Hurricane diving through a cloud layer.

types will come into their own. One can cite one phase in the Norwegian campaign as a case in point. For the short time during which it was possible for them to do so a squadron of Gladiators did sterling work. No other modern fighter type could have functioned from the "aerodrome" concerned.

Very often before I have remarked that the best thing to be said about any aeroplane is that it is easy to fly, and that the present British fighting types have that virtue—the Hurricane, particularly, among them. When I first left the ground in one of these machines—with a due and inevitable feeling of incompetence and personal danger—I had previously put in a bare three hours on types which might reasonably be considered as fighter trainers, the Harvard and the Battle. The former to teach me something about undercarriages, flaps and c.s. airscrews, and the latter to give me an idea of the feeling of a thousand horse-power, and to make the broad, long, tabletop cowling with its exhaust stubs on either side a little less terrifying.

I confess that my first impressions of the Hurricane were of dust storms, acceleration and violent rudder corrections. Before I had properly gathered my wits the machine was in the air and I was fumbling with the U/C retraction gear and its safety catch. The dust storm was the result of leaving the hood open according to instruc-

FOR YUGOSLAVIA : A Hawker test pilot putting a Yugoslav Hurricane through its paces.

particularly terrifying in forced landings. If you have some engine left and if there are any good fields to be found in these days of projected parachute troops, you find one and carry out a normal landing. If the engine stops you just leave the undercarriage up and make the best of a bad job in any sort of open space which is facing one as the machine is turned into wind. Apart from the incidence of houses, chimneys, trees and telegraph posts a modern machine with its undercarriage up will skate and crash through almost anything without turning over and generally without damage to the pilot, while the old-fashioned affair with a spindly fixed undercarriage almost invariably turned over unless the chosen field was well-nigh perfect. If we return to fixed undercarriages for certain types—and there is a lot to be said for them—these must be instantly detachable in emergency. In fact, the idea might be applied to all types, though instantaneousness is important because it might only be discovered at the last minute that a certain chosen field is unsuitable.

Cockpit Layout

Although it is necessarily "bitty" according to present-day standards, the layout of the Hurricane is a good one, with most of the items where you would expect to find them. In older types there was a certain inconvenience in undercarriage and flap operation, since after selection through the gate it was necessary to hold down a "power" lever until the operation had been completed—which involved flying with one's left hand and leaving the throttle and airscrew controls, not to mention the trimming wheel, to their own devices for a matter of ten seconds at rather critical periods. Nowadays mere selection produces the results—the gate is on the right side of and below the dashboard with the flap indicator near it.

One curiosity about the Hurricane is that the ailerons do not stiffen up as one would expect them to do when the speed increases. They remain quite reasonably light and very potent, at least up to the sort of speeds at which one normally makes use of them. Yet at no time are they too light and they still continue to work down at the bottom end of the scale. The normal approach speed seems to be about 90 m.p.h., but one can safely come in a great deal more slowly than that if engine is used right down to the ground and during the holding-off process. All you really need is sufficient speed—or power—to give time for a controlled hold-off and to prevent the possibility of a stall and heavy wing-down landing while the machine's attitude is being changed from a steep glide to a fairly pronounced tail-down position on the ground. With a straight approach there is really no need for much speed with any machine, and in giving oneself plenty of margin one is only preparing for the worst and for the forced-stall effect of vigorous elevator movement at the last moment. Hence, largely, the rumble approach.

The most important features of the Hurricane as an operational type are probably the good view from the pilot's seat and the natural sturdiness on the ground. Even during the process of landing, the ground can always be seen, and the Hurricane will take rough handling, both when landing and taxying, better than most. Hence its particular value as a front-line fighter operating from fields which have been torn up by bombs and gunfire, or which are, in any case, small and badly surfaced. A great many pilots will be unhappy if the Hurricane ever goes out of service, even in the distant future.   INDICATOR.

FRIEND or FOE?

Identification of Aircraft Simplified : A New Series of Articles

Hawker Hurricane, Single-seater Fighter

Wheels retract inwards, exhaust stubs above centre line. Radiator-scoop, beneath fuselage, in line with cockpit. Rounded wing-tips and fillets at the wing roots. Unbraced tail-plane. Large fin and rudder.

IN this new series Flight aims at helping you to identify hostile aircraft quickly and surely. British and American types will be compared with German and Italian. In the main, "opposite numbers" which resemble each other at first glance will be paired off and their recognisable differences made clear. We begin this week with the Hawker Hurricane and the Messerschmitt ME 109.

Messerschmitt ME 109, Single-seater Fighter

Wheels retract outwards. Exhaust below centre - line. Radiator beneath engine. Square-cut wing-tips, no fillets. Braced tail, small fin and rudder. Two small scoops, one behind each wheel in wing centre-section.

PROMPT identification of aircraft is of paramount importance to-day. Many thousands of persons in the R.A.F., the A.A. and searchlight batteries and the Observer Corps and Home Guard, in addition to thousands of roof-spotters, A.R.P. personnel, and even ordinary civilians, are learning to recognise friend from foe when an aircraft comes into view. In planning this new series of articles and illustrations for guidance, the primary fact has been borne in mind that the watcher invariably has only a few seconds in which to make his diagnosis. It is imperative that he should be able to identify an approaching machine quickly, and equally important that he should make no mistake in nationality.

For this reason it is intended in this series of articles to approach the whole question of identification from the point of view of the "spotter," who may get a near or distant view of his quarry, rather than from the purely technical aspect when the two angles are not wholly parallel. For example, the Junkers Ju 88 dive-bomber is fitted with in-line engines (inverted Vee liquid-cooled) but they have circular nose radiators which give the appearance of radial engines as employed on most British twin-engined aircraft. The man on the ground, therefore, whose duty it is to identify the aircraft as friend or foe, will only be concerned with its external appearance; what sort of engine happens to be hidden within the circular nacelles and the armament carried need not concern him, and to burden him with information which he does not require would only be to complicate matters unnecessarily for him. Brief details of power-units, together with overall dimensions, are already given on the *Flight* charts for the benefit of many who are interested in technical details, but such information is regarded as outside the scope of

these notes which will concentrate entirely upon externals and their ready recognition.

We begin to-day with a comparison of the Hurricane and the Me 109 single-engined, single-seater fighters. When seen approaching head-on in the distance it will be almost impossible to differentiate between these two opposing aircraft because each will show the bulge of a radiator scoop below the fuselage. As the machine approaches, however, the watcher may also be able to see—assuming good visibility—a small bulge under the centre-section of each wing; if so the machine is an Me 109. In conditions of imperfect visibility, or if the approaching machine is viewed from an angle, these small scoops, or ducts, may not be discernible, but if anything like a side view is presented, notice the "fore and aft" position of the radiator scoop. On the Hurricane it is behind the inwardly retracted wheels, but on the Me 109 it is immediately beneath the engine and in front of the leading-edge of the wings. Now note the position of the exhaust-stubs. On the Hurricane (as also on the Spitfire, Defiant and Battle) these are above the centre line, but on the German machines they are well below the line of the propeller-shaft.

When seen from beneath, the square-cut wing-tips of the Me 109 will easily be recognised even if other details, such as the braced tail, are not clearly visible.

It will be noticed that in the above picture of the Hurricane the flaps are shown in the down position, although the wheels are retracted. This was done to give a clear initial example of the proportions of the former and, at the same time, illustrate the appearance of the inwardly retracting undercarriage as it would be seen in normal flight.

HURRICANE	ME 109

THE NEW "WAR BIRDS"

A NEW DAWN.—American pilots are now flying Hawker Hurricanes in the Eagle Squadron of the Royal Air Force. These citizens of the United States who are taking a front line part in the battle for freedom are worthy followers of their compatriots who joined the Royal Flying Corps in 1917 and whose deeds are immortalised in the classic "War Birds."

THE AMERICAN Eagle Squadron is now an operational unit of the Royal Air Force. It completed its training and went on Active Service as a fighter squadron on Feb. 1 equipped with Hurricanes. At the end of six weeks it had had no opportunity of testing its prowess in combat with the Luftwaffe, but had taken its full share of patrols in the area to which it was allotted. Recently the squadron was At Home to a party of Press representatives—mostly Americans—at its operational station.

Early in its existence this squadron was the victim of that peculiar form of popular journalism which ignores essentials and arrays insignificant trifles in magnificent garb on the plea that stories must have "human interest." Some of the cables sent to America after a Press visit to the squadron when it was training did a grave disservice to a fine body of men, and held them up to ridicule and contempt. Excessive passion for "human interest" made it seem that the members of the Eagle Squadron were resolved to out-Lafayette the American Lafayette squadron of the last War, and that they were training more for a life of social conquests than for skill in air combat.

If that impression persists in America, or elsewhere, it can be dismissed and forgotten. The men of the Eagle Squadron have precisely the same virtues and vices, the same merits and faults, as the men of any other R.A.F. squadron. They have the same insatiable love of flying, the same confidence in themselves and in their machines, and much of it comes from their long and careful training and the preservation of an alert mind in a sound body.

In the Air

Their guests at the operational station watched some of them in the air. Six were to have flown but a slight mishap to one of the Hurricanes as it taxied out caused a temporary delay and the others took off without it. They rearranged their plans in the air by radio telephony, and the precision with which they took up and held their improvised formations dispelled all doubts about their skill or their steadiness. Holding a Hurricane in tight formation alongside another, low down and at speed, calls for more than courage or dare-devilry, yet at times the two outside men each seemed to have a wing-tip resting almost on the leader's cockpit. In individual flying, the five pilots showed themselves masters of the clean, smooth climb and turn on a wide open throttle—manoeuvres that were performed in a sky turbulent with bumpy air as the sun strove to disperse a thick morning haze.

In a week or so the squadron will start training in night operations. Meanwhile, it continues to perfect its fighting efficiency by day, using as its main instrument the camera-gun, which photographs and records the quality of the pilots' gunnery. The visitors were shown films taken during sham battles, and were surprised at the inability of some of the "enemy," despite violent twistings and turnings, to escape from the lenses, which, in real battle, would have been the guns of their pursuers. The camera-gun is worked by the gun button on the control column and operates only when, in more warlike circumstances, the eight Browning machine-guns would be in action. When the films are developed they are projected on to a screen and the pilots have visual evidence of their marksmanship. Errors can be observed and corrected; mistakes in tactics pointed out.

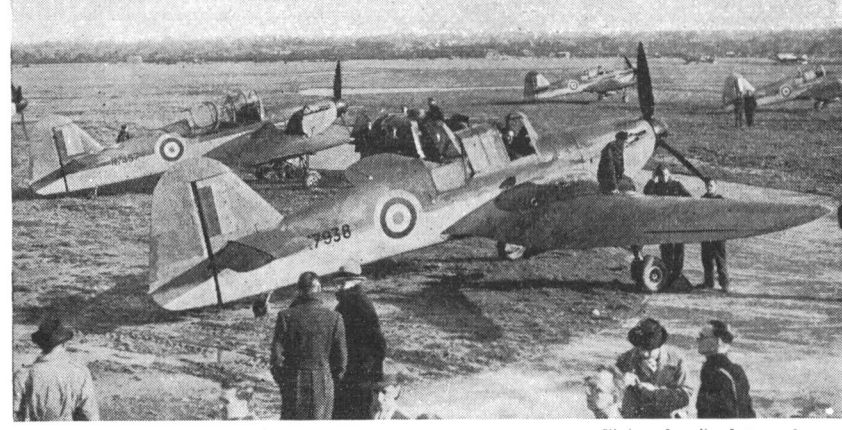

[" Aeroplane" photograph

SCHOOLDAYS.—Pilots of the Eagle Squadron graduated on Miles Master advanced trainers before undertaking operational duties on Hawker Hurricanes. This photograph was taken when the squadron was nearing the end of its training.

On other occasions the squadron " mixes it " with a nearby squadron of Hampden bombers, to the mutual benefit of both. The Hampden, rid of its bombs, has a surprising agility and is an excellent medium for putting a fighter through its paces. The Hurricane, being a nimble fighter, is an excellent medium for teaching bombers how to evade attack.

The Pilots

The squadron's guests had the pleasure of meeting some of the pilots who were off duty. When, earlier, the Commanding Officer had said that his " boys " were reticent and unassuming, he was not speaking with his tongue in his cheek. None wanted to talk about themselves, to give reasons why they came to Europe to join in the War, or to discuss their previous careers. But those visitors who were ready to talk to them about flying found them not only responsive but eager to discuss aeroplanes and equipment as long as time allowed.

This can easily be understood. Candidates for the Eagle Squadron must have at least 250 hours of solo flying before they can be considered for membership. All those now serving meet this condition with plenty to spare. Some have qualified on small machines like the Taylor Cub, but the majority have logged their hours on a wide range of types. One had been, among other things, an instructor at a flying school, a private owner and pilot to a rich Mexican ranch owner. He was also one of the first to come to England, and has flown the Buffalo and Spitfire as well as the Master and Hurricane and smaller types since he arrived. Another member of the squadron has had the rare distinction of flying a Tomahawk in England.

Their varied experience makes their views on aeroplanes and equipment worth listening to. They can compare, from first-hand knowledge, the merits and defects of both British and American systems, practices and methods, and their praise and condemnation is bestowed with complete impartiality.

Squadron Leader W. E. G. Taylor, the Commanding Officer, is typical of the whole squadron. He is quiet in manner and speech, shuns publicity and wants only to be allowed to " get on with the job." He would strongly resent any suggestion that he has special qualifications, heroic or otherwise, but he has undoubtedly had a useful, if unspectacular, flying career. He joined the U.S. Navy and learned to fly at Pensacola, the U.S. naval air base in Florida. He went to the Fifth Fighting Squadron when it was attached to the U.S. Aircraft Carrier Lexington. Afterwards he went back to Pensacola as an instructor and was later transferred to the U.S. Marines as a Reserve Officer. For six months he piloted air liners on the New York-Chicago route to sample bad-weather flying.

He came to England in August, 1939, resigned from the U.S. Marines and joined the Fleet Air Arm when War broke out. He served with a Naval fighter squadron in the defence of Scapa Flow and on the Aircraft Carriers Glorious, Argus and Furious. He was transferred from the Fleet Air Arm to the Royal Air Force to command the Eagle Squadron on its formation last October. He does not claim to have shot down any enemy machines so far, though he had at least one encounter with the Luftwaffe which might have ended in his favour. " But," explained Squadron Leader Taylor, " Jerry was still going strong when we last saw him."

Guidance From the R.A.F.

Four R.A.F. pilots were loaned to the Eagle Squadron to supervise its operational training. They had fought through the hottest period of the Battle of Britain last Autumn and had no need of textbooks from which to extract their knowledge

ON ACTIVE SERVICE.—*A pilot of the Eagle squadron on the wing of a Hawker Hurricane fighter bearing the squadron's crest. The squadron became an Operational Unit on Feb. 1 this year, after four months' training.*

of fighting tactics. These pilots will ultimately be replaced in the squadron by Americans, but until they are they will continue to serve as Flight Commanders. The Americans may have found some difficulty in getting together as a team—not because they are unamenable to discipline, but because they were all seasoned pilots accustomed to flying more or less in their own sweet way.

They know that the War has thrown up several outstanding fighter pilots, but experience has shown that the squadron that can fight well as a team is likely to win a far better record in the long run than one made up of brilliant individualists. This fact is sometimes difficult to impress upon pilots who possess a particular flair for showmanship, and it certainly goes hard with those who recall the deeds of famous airmen of the last War and aim to emulate them.

All the pilots of the Eagle Squadron hold commissions. No one knows why this is, but the tradition seems firmly established and likely to be perpetuated.

Pilots from America are reaching this country at the rate of 30 a month. Already the Eagle Squadron can accept no more, but whether new Eagle Squadrons will be formed or not seems still undecided. The newcomers may be absorbed in other R.A.F. squadrons and the Eagle may remain the representative American squadron with the Royal Air Force throughout the War.

The original plan was to revive the Lafayette Squadron, which, in the last War, was raised by Americans in France and attached to the French Air Force. It was re-formed in 1925 to fight for France against the Riffs in Morocco. When France collapsed last year and the Lafayette scheme fell through, its sponsors, Mr. Charles Sweeny, an American business man living in London and now Liaison and Reception Officer of the Eagle Squadron, and his uncle, Group Captain Charles Sweeny, proposed the formation of the Eagle Squadron of the Royal Air Force.

Thus, history repeats itself. In the last war Americans served in their own squadron with the French Air Force and as individual pilots with squadrons of the Royal Flying Corps. The exploits of the Escadrille Lafayette were described in the book " One Man's War," by Bert Hall; those of the Americans in the R.F.C. in " War Birds," which was published as " the diary of an unknown aviator," and was one of the best books the war in the air produced.

Pilots of the Eagle Squadron who have read either of these books will have learned already the difference in conditions between this War and the last. Regimentation has come to supplant improvisation, the team to replace the individual, and qualities beyond those of sheer courage and skill are now demanded from those who fly in operations against the enemy.

To what degree the Eagle Squadron has them has yet to be proved, but the visitors who recently saw the pilots on their station could not help sharing the opinion of the Station Commander when he said, " Now the Squadron is operational and we are likely to hear big things about it."

TWO'S COMPANY.—*A section of the Eagle squadron coming in low over another Hurricane of the same squadron. Later, five of the pilots flew in formation and earned full marks for their station keeping.*

FRIEND or FOE ?

Tail Unit Design as an Aid to Identification

(Continuing a regular weekly series which commenced in *Flight* of January 9th, 1941, and incorporated 33 comparative studies in detailed points of identification)

HAWKER HURRICANE. Cantilever tailplane set just below base of fin. Rudder projects between elevators and extends down to bottom line of fuselage. Navigation light on rudder. Curved outline to all surfaces.

THIS week begins a new phase of the " Friend or Foe? " identification series. The first phase, just concluded, presented a total of 66 different types of military aircraft during 33 consecutive weeks. These were illustrated by three-quarter front views, accompanied by sets of general arrangement drawings showing front and side elevations and the plan from beneath. Identification of aircraft continues to be a subject of first importance and general interest, and *Flight's* correspondence files bear ample witness to the usefulness of its " Friend or Foe? " pages.

The new phase will continue to present current types from a different angle and one which brings tail design into prominence. Quite often a machine can be more readily identified by its empennage than by any other single feature, and it is easy to imagine circumstances in which the spotter may have to depend upon recognising a characteristic tail design in order to confirm identity. For that reason the normal G.A. drawings will be replaced by plan and elevation drawings of the tail-unit alone.

We begin this week with the Hawker Hurricane and Heinkel He 113. Both these tail groups are on the same general lines; that is, the tailplanes are cantilever and are mounted just below the base of the fins, and the rudders extend down between the elevators to the bottom line of the fuselages. But they can be distinguished from each other at a glance by the fact that the tail outlines of the British machine present a series of graceful curves, while those of the German fin and rudder are far more angular. This difference, while certainly a national characteristic in design, must not be regarded as anything like an infallible clue to nationality, for there are a number of exceptions on both sides, as will be seen in subsequent examples in this series.

Hurricane

He 113

All that can safely be said on this point, therefore, is that any machine seen to possess an angular outline to its tail assembly should at once be suspected of hostile origin until the spotter has had sufficient chance to identify it with certainty. This only applies, to any appreciable extent, to machines seen over this country or its adjacent waters ; in the Middle East and Mediterranean theatres of war, where Italian aircraft are occasionally encountered (when a rear view is particularly likely!), the more artistic Latin temperament seldom indulges in aerial angularity.

Certain detail differences will be observed from a study of the accompanying pictures. The Hurricane, for example, has a navigation light fitted to the extremity of its rudder and immediately below this is an inset trimming tab ; each elevator, however, has a projecting tab. Probably because it is *never* safe for a German machine to display navigation lights at night—not even over its own territory, these days—no tail lamp adorns the rear of the Heinkel. In the fitting of trimming tabs the reverse method has been adopted on the German fighter, that on the rudder projecting and those on the elevators being inset on each side of the " bite." Yet another divergence is in tail-wheel design. That of the Hurricane is fixed (it retracted on earlier models), but a shallow fin, which acts to some extent as fairing, runs along the bottom of the fuselage from just aft of amidships to the base of the rudder post. The He 113 tail-wheel retracts completely.

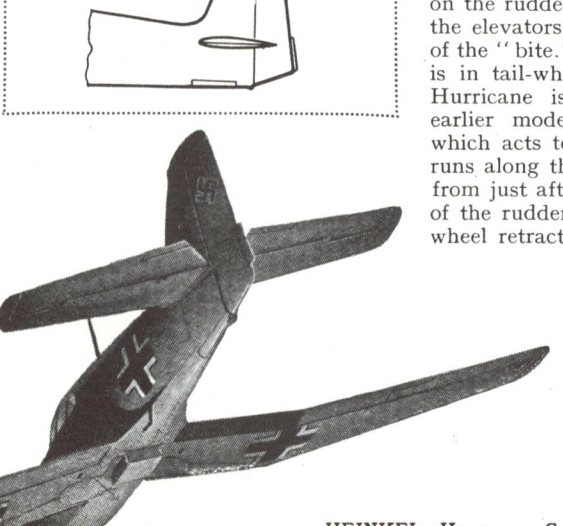

HEINKEL He 113. General layout of the tail group is very similar to that of the Hurricane, but outlines suggest a series of straight lines and sharp corners instead of curves.

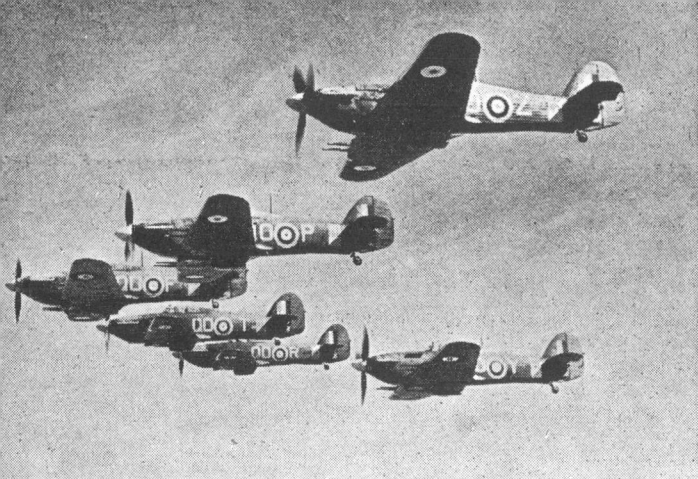
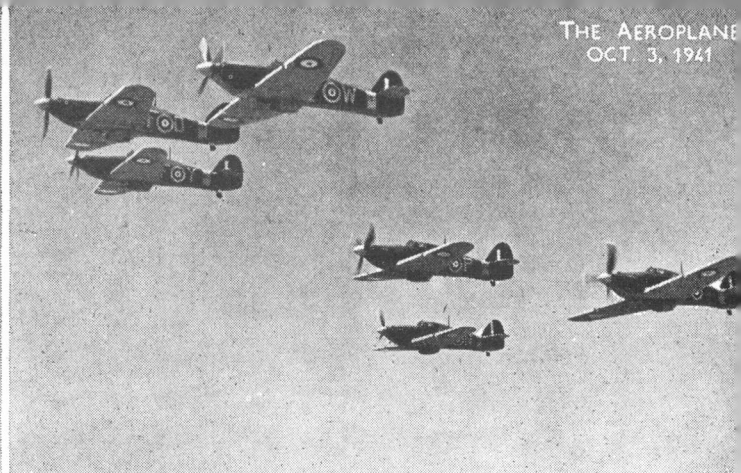

The Hawker Hurricane II

THE World's greatest fighter re-vitalised—the Hawker Hurricane II—has a newer version of the Rolls-Royce Merlin motor and the phenomenal armament of either 12 machine-guns or four cannon.

As our chief defender in the Battle of Britain the Hurricane earned a place in history such as few of man's mechanical creations can claim. That victory followed on a short but heroic career in Norway and France and led on to even greater achievements in Libya, Eritrea, Abyssinia, Malta, Greece, Syria, Iraq and now Russia.

Such a series of campaigns might well have exhausted the usefulness of even the greatest of aeroplanes. In fact, nothing is farther from the truth, and through the untiring work of the two companies which have shared its development—Hawker Aircraft Ltd., and Rolls-Royce Ltd.—the Hurricane of to-day gives promise of more years of historic achievement.

Designed originally by Sydney Camm in 1934, the Hurricane had already undergone many modifications and improvements when it took part in the Battle of Britain last year. Metal wings had replaced fabric covering, and the fixed-pitch airscrew had given place to a constant-speed airscrew driven by an improved Merlin motor. That was the Hurricane I.

The Hurricane II of to-day is as great an advance on last year's model as that on the original version. The Rolls-Royce Merlin motor has been stepped up in power again so that now the rate of climb is quite extraordinary and the ceiling outstanding.

More noteworthy still, the armament has been drastically revised and now consists of twelve 0.303-in. Browning machine-guns (Hurricane IIB) or four 20 mm. British Oërlikon cannon (Hurricane IIc) The former have a total rate of fire of some 14,400 rounds per minute giving a weight of fire of about 360 lb. per minute.

The Hurricane IIc with four 20 mm. cannon has a greater fire power (600 lb. of explosive shell per minute) than any other single-motor fighter in service anywhere.

Both types of armament are in use because opinion is still divided as to which is the more effective. To a certain extent the choice depends on the duties to be performed. The cannon in the Mark IIc slightly reduce the maximum speed.

DIMENSIONS.—Span, 40 ft. 0 ins; length, 31 ft. 9 ins.; height, 13 ft. 1½ ins.; wing area, 257.5 sq. ft.; aspect ratio, 6.2.

WEIGHT.—Loaded, more than 7,000 lb.

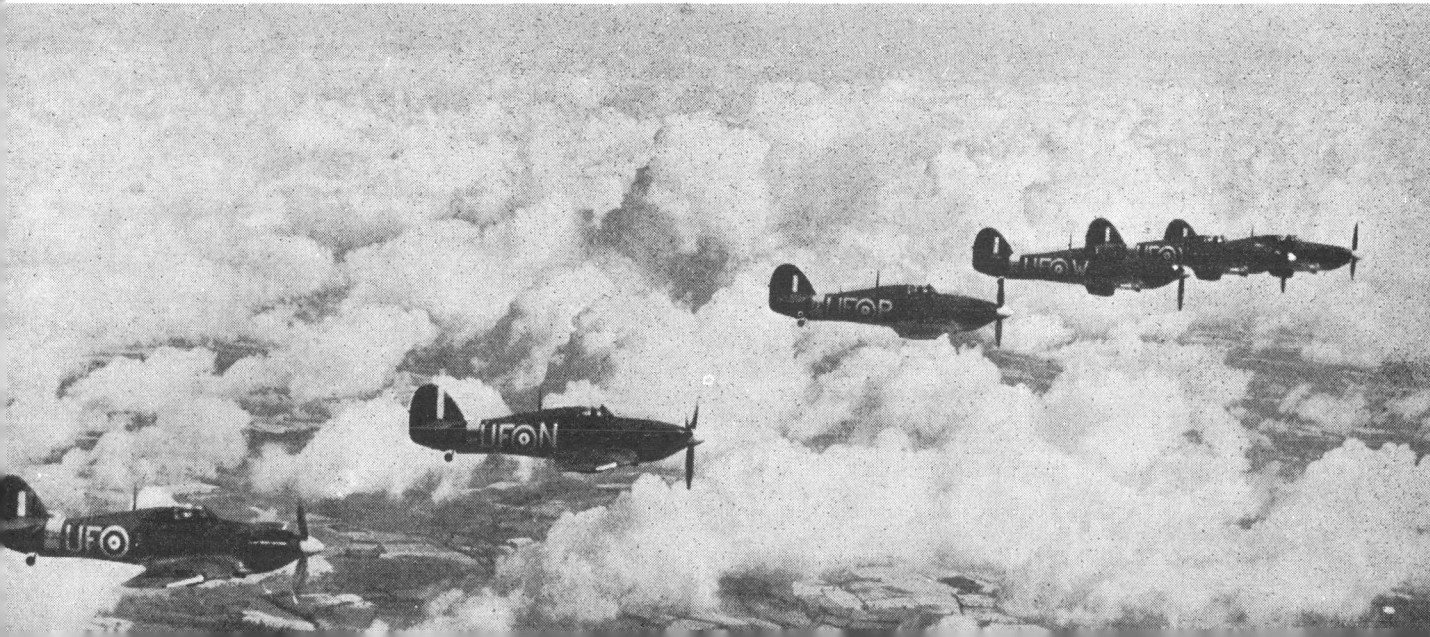

FIGHTER-BOMBER
Squadron

Hurricane bomber silhouetted against cumulus cloud.

"Flight" Visits a Squadron of the R.C.A.F. which Flies Hurricane Bombers : Ground Defences go Up to Shoot Down

MR. CHURCHILL, when reviewing the war situation recently, recalled that in the last war, when every possible shortage had been predicted, all that we actually ran short of was Huns. To-day some of our Hurricane bomber squadrons are already in that predicament. The occupied coast aerodromes from which the *Luftwaffe* issued to be slaughtered in the Battle of Britain last year are now but advanced landing grounds held by a minimum of personnel. The coast roads on which last spring the German army tramped daily to and from their invasion practices are now deserted of military traffic. If a ground-strafing Hurricane pilot is lucky enough to find a lorry full of German soldiers on which to train his battery of guns he immediately becomes the envy of his squadron. Doubtless, a certain amount of the absence of garrison is due to the necessity to make good some of the losses on the Russian front, but the daylight sweeps of Fighter Command have had a big effect.

Of all the many excitements which may fall to the lot of an R.A.F. pilot in wartime—and there are plenty—there is just nothing to touch this ground strafing for sheer thrill. Merlin roaring flat out, A.S.I. showing over 300 m.p.h., over a hedge, round a tree, under or over a high-tension cable as opportunity affords, aim between two factory chimneys, a last jink to put the ground gunners off. Now the Brownings blaze with a noise like ripping calico as the thumb presses the gun trigger on the joy stick. Steadying for a fraction of a second while the bombs are released and then away into any convenient cloud for cover or more hedge hopping until out of range once more. A few minutes' madness and it's all over. Formation is picked up on the leader and course set for home at an economical cruising speed.

Recently we visited one of the new Hurricane bomber stations and were able to chat with the pilots about some of their experiences on the other side. Unfortunately, as

The 250 lb. bombs fit snugly to the streamline racks. They are suspended from that portion of the wing which is already strengthened to take the guns.

A Hurricane flies low past the camera.   All the guns and bombs are carried clear of the slipstream.

is the case with all war reporting, the most interesting parts of the proceedings are secret, and it would be more than foolish to help the Hun in the slightest degree.

The pilots who were our hosts belonged to a squadron of the Royal Canadian Air Force who arrived in this country at about the lowest ebb of our fortunes. They have been on operations for some while now.   In the operation from the evening of November 7th until the evening of the following day—which constituted the greatest day of the R.A.F. to date—they supplied a number of the 300 aircraft which Fighter Command sent over the Continent during those twenty-four hours. The Hurricane bombers, with three squadrons of Spitfires as an escort to look after the upper air, swept in with only the difference between the height of their own flying field and that of the French cliffs showing on their altimeters.   For a target they had been allocated a factory which was known to be on war work for the Germans. With hundreds of thousands of bullets squirting from the eight Hurricanes' guns to keep the ground defences' heads down their sixteen 250-lb. bombs were placed well and truly in the target area.   As they turned for home the short-time delayed-action bombs exploded and the whole factory was seen to burst into flames.

A target which is always popular with the pilots is the odd electricity

Canadian armourers re-arming a Hurricane while the pilot stands by in readiness.

In addition to the numerous opportunities for the real thing, exercises with 8½ lb. practice bombs are carried out frequently.

transformer. Apparently a good squirt of bullets into one is quite equal to the set-piece at any Brock's firework display. It seems almost a shame to spoil the show with a bomb As already indicated, these attacks are made from an extremely low level, using every fold in the ground to hide the machines until the final moment of attack. It is a fact that the Germans are building 30ft. A.A. towers in order that they may shoot *down* on to our Hurricanes as they zip across flying fields and other targets at over 500 ft./sec. We cannot imagine these towers are very popular with the crews who man them. This very low flying foxes every form of detection, and if it were not for the approach having to be made from the sea, complete surprise would be achieved more frequently than it is. What usually happens is that the shore guns open up while our pilots are still some three miles out to sea. The Germans have, on occasion, tried this ground-strafing with their Me 109 fighter-bombers, but they seem to prefer cloud cover for the approach.

Bombing from this height brings in its train problems of its own. To be reasonably safe from the explosion of an ordinary 250 lb. bomb, the aircraft must be at a greater height than 1,500ft. Over 2,000ft. is a good deal more comfortable. It is for this reason that short-fused, delayed-action bombs are used to prevent the pilots from blowing themselves up. For the same reason the attack is made roughly in line abreast. If line astern were used, the last man would stand an excellent chance of being blown up by the bombs dropped by the leader. Bomb ballistics also present a considerable problem. The bomb when it hits the ground from such a low height ricochets along in the horizontal position. More in the nature of a shell fired at point-blank range than the accepted idea of a nearly vertical trajectory. This effect is all to the good when fairly massive targets, such as power stations or marshalling yards, are attacked, but one pilot who watched another aircraft attack a railway station saw the bombs

go clean through both walls of the station and explode about 300 yards away in some fields. This level trajectory carries one distinct advantage in that no complicated bomb sights are required. When the target appears to be a nose length ahead of the spinner, the bombs are released.

It would be unfair to tell this story without giving the Hurricane a pat on the fuselage. As an eight-gun job with a two-bladed wooden airscrew, it was a good fighter in 1935. With a v.p. three-bladed airscrew in 1939 it was better. A more powerful Merlin to-day maintains a higher maximum speed for the Hurricane than the original type, despite the fitting of 50 per cent. more machine guns or four 20 mm. cannon, or a combination of machine guns and two 250 lb. bombs.

Pilots who fly the bomber version are quite confident that they could fit two 500-pounders without sacrificing much in the way of performance.

With the guns and bombs all carried outside the airscrew disc it would be expected that manoeuvrability would suffer somewhat. This, however, is not the case, and the pilots are quite definite on this point. Of course, if an unlucky bullet stops the engine, the stalling speed does go up a good deal with the added resistance of a dead airscrew.

The Allotted Span

SOME confusion has been caused by the discrepancy between the general arrangement drawing of the Avro Manchester in our issue of November 13th, 1941, and the printed measurement of 90ft. 1in. for the wing span.

The original official figure gave the Manchester a span of 80ft. A second official communication quoted this figure as 90ft. 1in. We are informed that this second measurement is the correct one. The length is 70ft. and the height 19ft. 6in.

Canadian Lancasters

THE Lancaster bomber which, as announced by Mr. Howe, Canadian Minister of Munitions, is to be built in Canada at the request of the British Government, is one of the latest of our growing fleet of four-engined bombers.

No details of its design have yet been disclosed in this country, but reports from a Canadian source state that it will be powered by four Bristol Hercules engines and that not only will it carry a very great bomb load, but it will be one of the fastest four-engined aircraft ever designed.

Following the famous Hurricane I, our chief defender in the Battle of Britain, the Hawker Hurricane II has a more powerful Rolls-Royce Merlin engine and the amazing alternative armament of four cannon or 12 machine guns. The photograph shows a formation of Hurricane IIC's on patrol.

Advt.

ARMY CO-OPERATION IN

A Photographic Survey of the Flying, Living and Working Conditions of a Hurricane Squadron Giving the Eighth Army Close Support

1

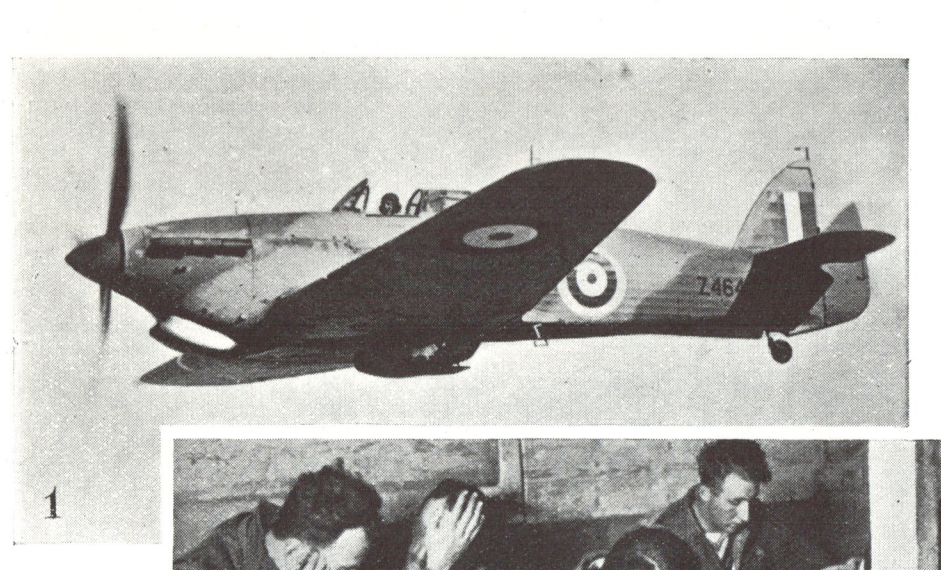

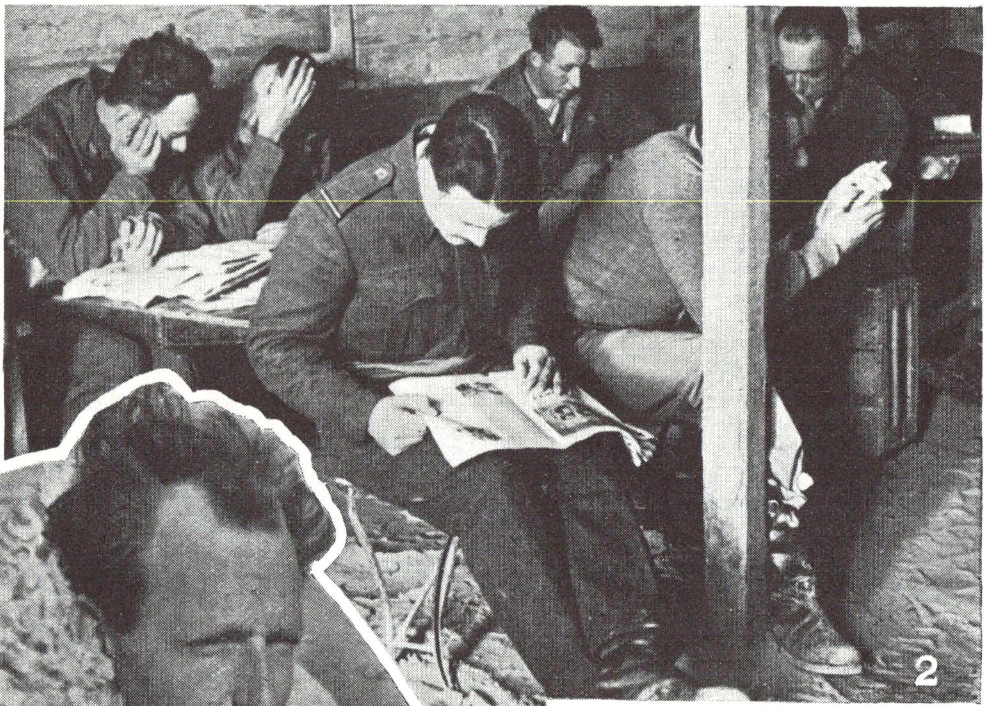

2

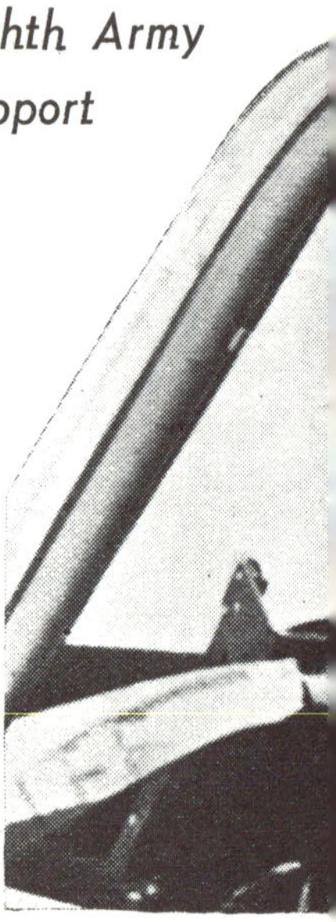

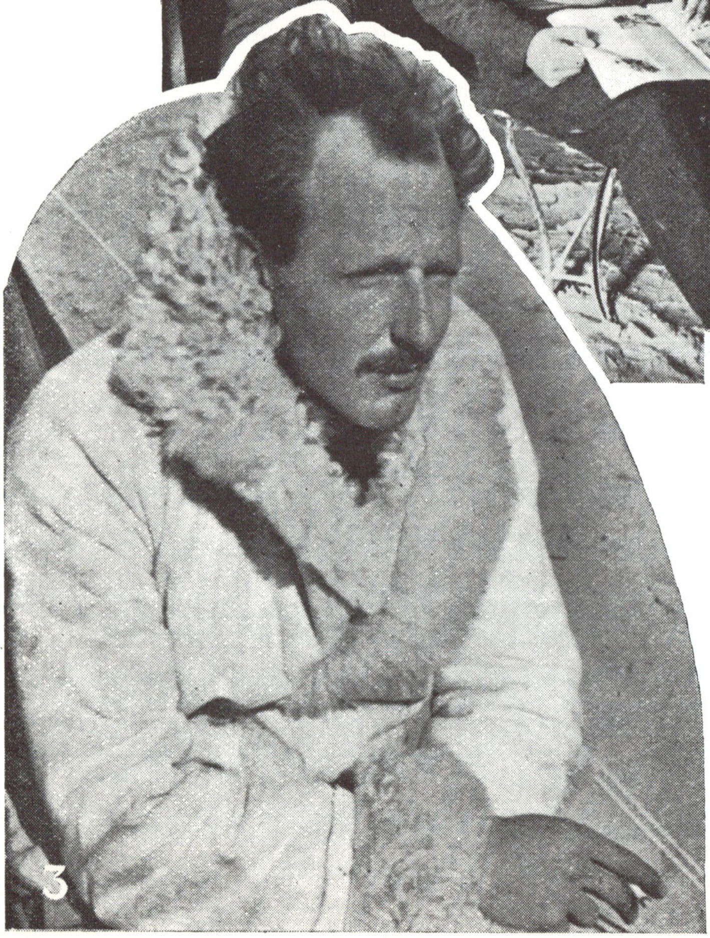

3

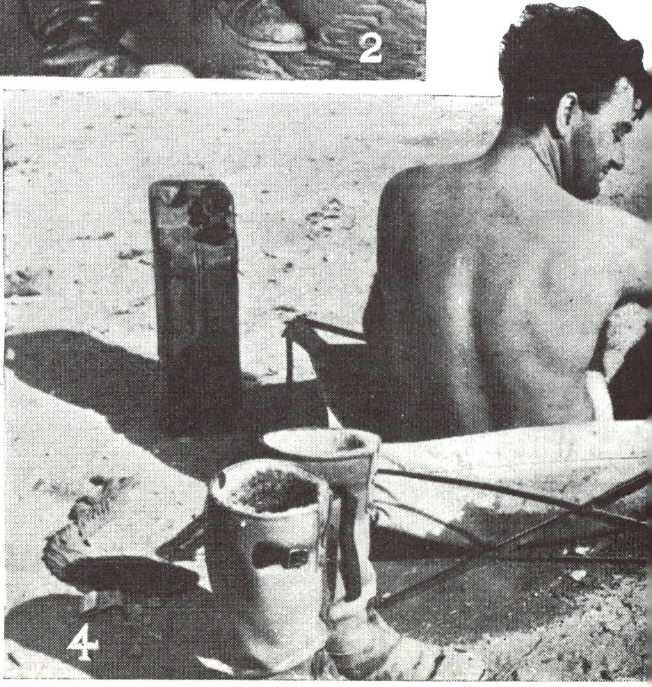

4

LIBYA

(1) A Hurricane off for an artillery shoot. Note the special air cleaner intake. (2) "Comfort" in the officers' mess. (3) A sheepskin to keep out bitter winds. (4) If one is clever one can be clean. (5) Pilot wearing the combined microphone and oxygen mask which is standard in the R.A.F. (6) A fire for warmth in a desert which the popular mind thinks is always scorched. (7) The army liaison officer discusses operations with the pilots. (8) The radio ground station talking to the air.

AIRCRAFT TYPES AND

Hurricane IIB

ORIGINALLY designed as a single-seater, eight-gun fighter as long ago as 1935, the Hurricane has proved one of our most versatile and adaptable aircraft and the Mark IIB differs only from the Mark I in the matter of certain vital equipment.

Structurally, the design which proved itself in France, Norway and Libya remains basically unchanged, the fuselage being composed of a rectangular, rigidly braced primary structure of steel and aluminium alloy tubing, and a secondary structure of light wooden formers which fair the outline to an oval section, fabric covered. The wings, however, are an all-metal structure comprising a rectangular centre-section accommodating the inwardly retracting undercarriage, and two tapered outer panels, the whole being stressed-skin covered, flush-riveted over the leading-edge. Ailerons are fabric covered as also are the surfaces of the metal-framed tail unit except for metal-covered leading edges.

The first of the Mark II series had the Rolls-Royce Merlin III liquid-cooled V-type 12-cylinder engine, but later models are powered by the Merlin XX with two-speed supercharger. The former engine developed a maximum power of 1,030 h.p. at 3,000 r.p.m. at 16,250ft., and gave the Hurricane a top speed of 355 m.p.h. at 18,500ft. and a service ceiling of 35,000ft. The Merlin XX, however, gives a maximum power with the low-gear supercharge of 1,260 h.p. at 12,250ft., and a maximum with the high gear of 1,175 h.p. at 21,000ft., with a useful increase in the performance figures, which, however, it would be imprudent to divulge.

Chief function of the Mark IIB is that of low-level attack and, popularly known as the Hurri-bomber, its armament comprises both machine guns in the wings and externally housed bombs beneath the wings, just outside the centre-section. Generally speaking, it carries two 250 lb. bombs for this purpose, but the bomb load can be varied and the number of guns varies from ten to twelve accordingly.

Another version of the Hurricane is the Mark IIC, which differs only in being equipped with four 20 mm. cannon —two in each wing—in place of machine guns. This type is easy to recognise because the cannon project boldly from the leading-edge, but all Hurricanes are characterised by their wing plan, depth of fuselage with its "hog's back," and the size and shape of the fin and rudder.

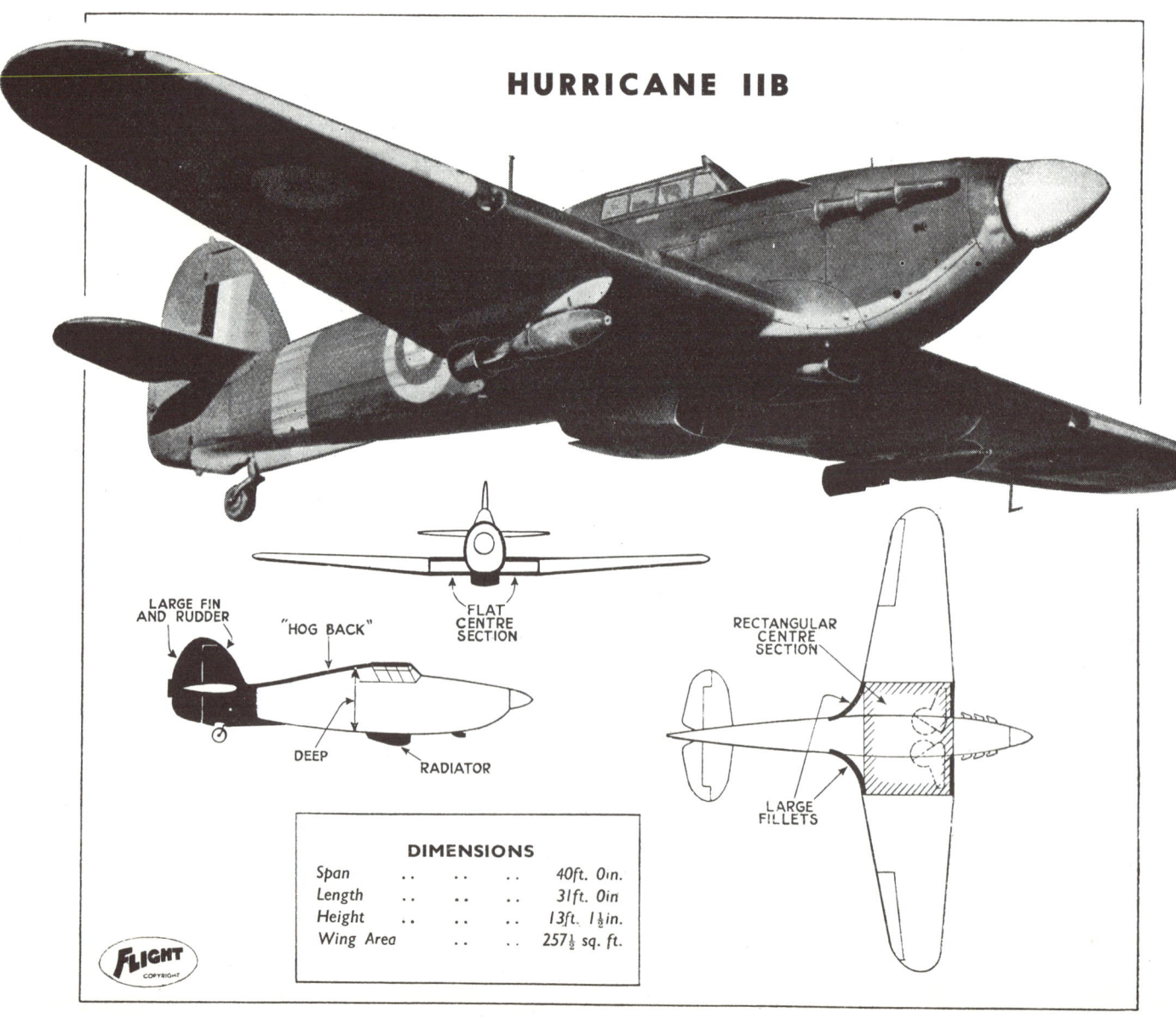

HURRICANE IIB

LARGE FIN AND RUDDER

"HOG BACK"

FLAT CENTRE SECTION

DEEP

RADIATOR

RECTANGULAR CENTRE SECTION

LARGE FILLETS

DIMENSIONS

Span	..	..	..	40ft. 0in.
Length	..	..	..	31ft. 0in
Height	..	..	..	13ft. 1½in.
Wing Area	..	..	..	257½ sq. ft.

THEIR CHARACTERISTICS

TROPICAL LONG-RANGE HURRICANE IIc

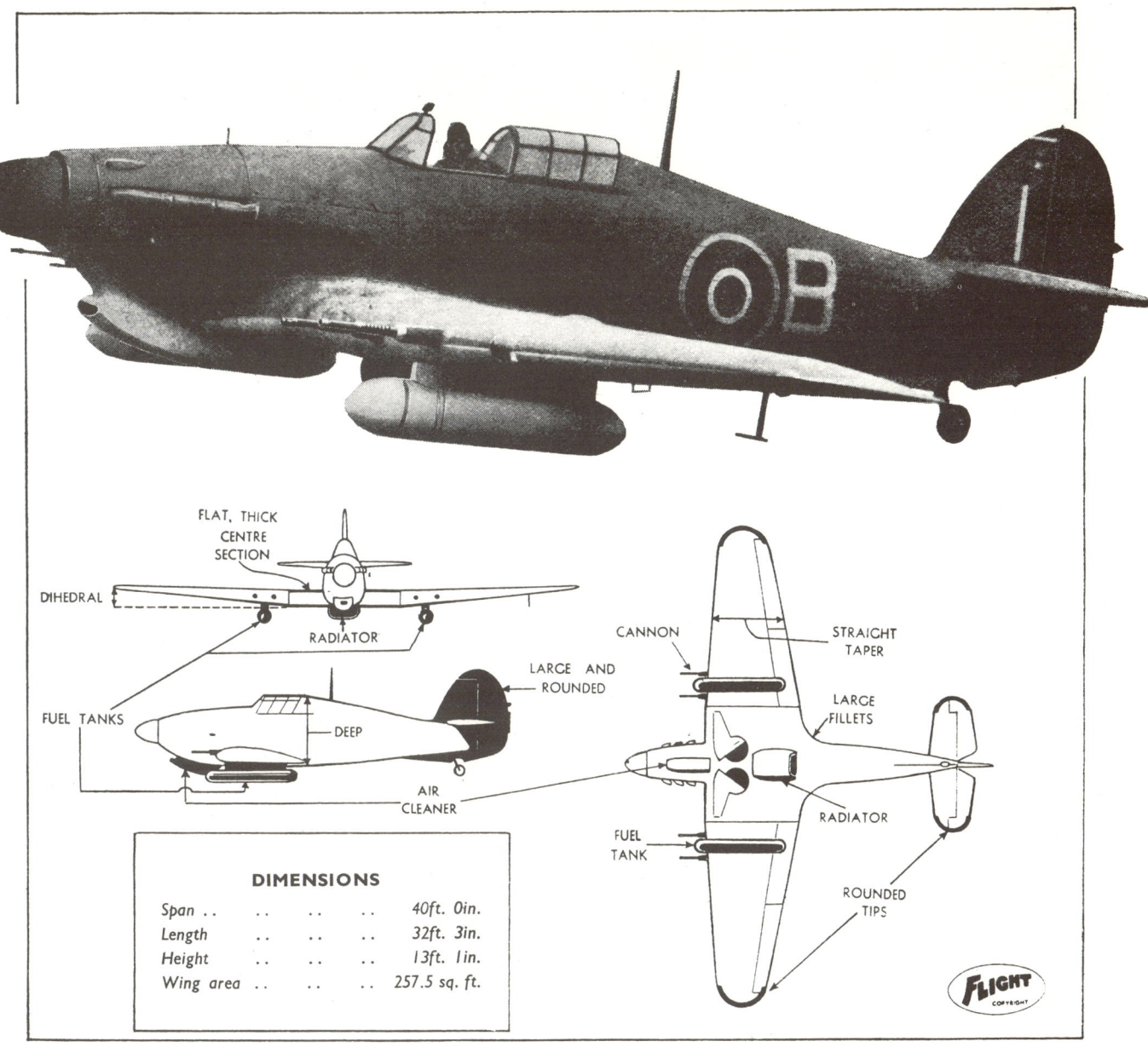

FLAT, THICK CENTRE SECTION

DIHEDRAL

RADIATOR

FUEL TANKS

DEEP

LARGE AND ROUNDED

AIR CLEANER

CANNON

STRAIGHT TAPER

LARGE FILLETS

FUEL TANK

RADIATOR

ROUNDED TIPS

DIMENSIONS

Span ..	..	..	..	40ft. 0in.
Length	..	..	..	32ft. 3in.
Height	..	..	..	13ft. 1in.
Wing area	..	..	..	257.5 sq. ft.

IT would be difficult to name an aircraft which has proved more adaptable and versatile than the Hawker Hurricane, nor one which has "kept its youth" to such good effect for so long a time. For the Hurricane is still a modern aircraft in spite of the fact that it first appeared more than seven years ago.

The variety of forms in which it has seen service at least equals that of the Spitfire in number, and certainly exceeds it in divergence, for it fulfils with admirable efficiency the roles of day and night fighter variously armed with machine guns, cannon, or a mixture of both, as a catapulted fighter from merchant ships, as a carrier-borne fighter with the Fleet Air Arm, and as a fighter-bomber. Furthermore it can be hung with auxiliary fuel tanks to increase its range and, like any other machine, be equipped with "tropical kit" externally manifested in the bulge of an air-cleaner for protection against desert dust.

It is in the last-named guise, namely the long-range, tropicalised version of the Hurricane IIC, that this R.A.F. maid-of-all-work is reviewed and illustrated on this page.

Some idea of the steadily mounting burdens placed upon the uncomplaining Hurricane may be gauged from the fact that its all-up weight has increased from less than 6,000 lb. to approximately 8,250 lb. for this latest version, yet the wing loading is still by no means excessive at 32 lb./sq. ft. As a result, its manœuvrability remains excellent.

Naturally one cannot pile on drag in the way eloquently illustrated above without some sacrifice in sheer speed, but even so, the tropicalised IIC with its two auxiliary long-range fuel tanks slung below the roots of the wing outer-panels, is still capable of a top speed of 312 m.p.h., which its 1,280 h.p. Rolls-Royce Merlin XX engine achieves at 18,000ft. Maximum cruising range is about 1,500 miles so equipped. Without the external fuel tanks and the tropical air-cleaner, its top speed is 340 m.p.h. at 21,500ft.

Apart from the special features already enumerated and shown in the photograph and drawings above, the basic Hurricane characteristics remain unchanged.

IN ITS LATEST FORM : Officially designated the Hurricane IID, this type with its two 40mm. cannons is more picturesquely known as the Tank Buster.

THE Hawker Hurricane has, without doubt, proved itself to be the most versatile weapon of this war. On these pages the Hurricane is shown in a number of its many types—from the prototype which flew in 1935 to the new 40 mm. cannon-armed tank attack IID model, which has played such a large part in the series of victories of the 8th Army.

Our story deals almost exclusively with the deeds of the Hurricanes in the Middle East Command wherein, over a long period, they have been fighters, bombers, Army co-operation and close Army support aircraft rolled into one.

At the same time, it must not be forgotten that the Hurricanes bore the brunt of the Battle of Britain nor that they were our standard " cat's-eye " night fighters for a while.

Hundreds of Hurricanes have also been delivered to the Russian Air Force, and the Russian pilots are very impressed by their fire power and flying qualities,

The Harassing Hurricane

Its Exploits in the African Theatre : Greece, Crete, Abyssinia, Somaliland, Eritrea

TAKE the skill, courage and devotion of the fighter pilots during the Battle of Britain. Take the determination to oppose the enemy, no matter what the odds and where the place. Take the perfect co-ordination of man and machine—and you have some assessment of the

The Hurricane IIC as employed in the Middle East. It has tropical radiator and air intake, two 20mm. Hispano cannons in each wing, long range tanks which are interchangeable with bombs and ejector exhaust stubs.

To protect our merchant convoys out of the range of our land based aircraft, a number of ships have been fitted with catapults to shoot off Mark I Hurricanes. When the fuel is exhausted the pilot either bales out or alights on the sea and is picked up.

THE HARASSING HURRICANE

part played by the Hurricane of Middle East Command in the driving of Rommel and his *Afrika Korps* behind the defences of Tunisia.

The Hurricane was the first modern fighter to appear in the Western Desert, and it was the first object of fear to the *Luftwaffe* and the *Regia Aeronautica*, every type of whose fighter and bomber in turn came under the Hurricane lash. There were many unequal battles in the early days, but by sheer offensiveness the R.A.F. Hurricanes won the aerial superiority that made the Eighth Army's advance possible. The greater the odds, the greater zest did our pilots show for the fight.

At one forward landing ground, five Hurricanes rose to engage a force of nearly 40 German and Italian bombers

Grp. Capt. P. W. S. Bulman, of the Hawker Aircraft Co., testing the prototype Hurricane in December, 1935. It had fabric wings, as did a number of the production models until the change over to all metal.

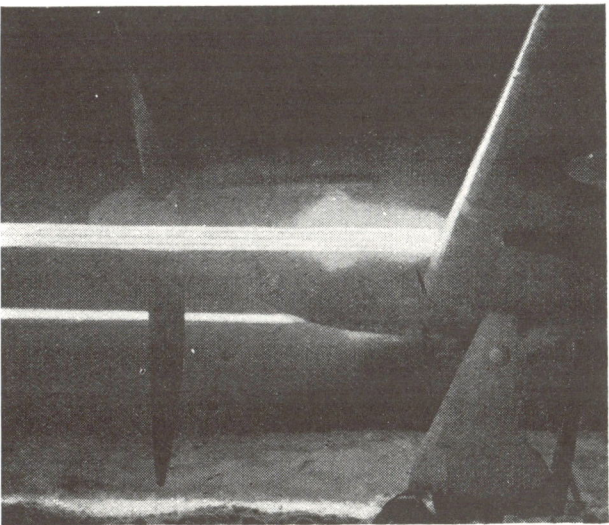

The eight .303in. Browning machine guns of a Mark I Hurricane being fired into the stop butts at night.

and fighters. Seven of the enemy aircraft were destroyed, with two probables. Several of the remainder were damaged before they headed for home.

There was a Durban pilot who went into his first engagement with the enemy undeterred by the fact that they were seven Stukas. He shot down two of them.

A formation of 20 German dive-bombers, accompanied by five Messerschmitts, was intercepted by 10 Hurricanes of a famous Australian squadron. Five dive-bombers and three Messerschmitts were shot down and the rest routed. Two days later, the same squadron destroyed 14 dive-bombers that ventured over without fighter escort. In these two engagements, the Australians lost only one fighter.

Low-flying Attacks

Axis ground forces also felt the weight of the Hurricanes' blows. Concentration of German and Italian troops, encampments, landing grounds, tanks, armoured cars, motor transport, armed motor-cyclist dispatch riders; all were fair game in fearless low-flying attacks.

One of the highlights of the British 1941 campaign was the exploit of a long-range Hurricane squadron which, for months, operated from an isolated desert base well south of the enemy forces. Bronzed, bearded pilots of this squadron harried supply lines, retreating columns and landing grounds far behind the fighting area. The squadron destroyed more than 30 enemy aircraft on the ground, 13 in the air, and more than 400 vehicles.

Few outside the gallant garrison of Tobruk realise the part played in that eight-months' stand by the Hurricane fighters who provided cover for the seaborne supplies to the port. In an effort to cut off these supplies, the enemy threw in dive-bombers and fighters in constant attacks on merchant ships and their naval escorts. Time and time again, patrolling Hurricanes beat them off and saw the precious cargoes safely to the beleaguered men.

Over the sea outside

The Hurricane IIB, the first of the bomber versions, carried two 250lb. bombs and had twelve .303in. machine guns in the wings. The engine was a Rolls-Royce Merlin XX.

THE HARASSING HURRICANE

Tobruk, a South African squadron shot down six enemy aircraft for the loss of one, and on another occasion a R.A.F. squadron accounted for five without loss.

Aircraft that attacked the town itself fared no better. One April day, nine German bombers appeared high over Tobruk. Hurricanes took off and shot down four of them. Suddenly, twenty escorting Me's attacked from above, and the troops on the ground cheered wildly when four of them were sent hurtling.

The Hurricane was the first fighter-bomber in the Western Desert, where it was operating against the enemy as early as March, 1941. As soon as it appeared, tanks, armoured cars and motor transport felt the weight of its bombs, followed by the just as effective power of its guns. The Hurricane jumped in as the R.A.F.'s desert man-of-all-work. Armed reconnaissance, Army co-operation, night fighting and photographic reconnaissance were only a few of the jobs the pilots took on, almost within hours of seeing the desert for the first time.

Greece—Crete : Hurricanes were in to the end. They chased a numerically superior Italian air force from the skies ; they operated for six months in some of the worst flying weather imaginable. They provided fighter cover for our forward troops until sheer weight of Nazi numbers forced them back to only two airfields in the Athens area. Pilots were given no rest, nor did they ask for any. Aircraft that were full of holes—in some cases with parts of wings blown off—went up again, without repairs, to cover the evacuation. Even when the airfields of Crete were denied them, long-range Hurricanes continued to engage

For low strafing and medium height fighting the Hurricane IIC with its four 20mm. cannon is a sturdy killer.

the enemy from bases in North Africa. The long sea crossings were covered without any thought of the personal safety of the pilots.

It was the same story in every African theatre of war : Abyssinia, Somaliland, Eritrea. It was the same story in Malta. There was a time when a handful of Hurricanes was the island's only fighter defence. These aircraft, and the boys who flew them, went through an ordeal that tested man and machine to a point, theoretically, beyond human and mechanical endurance. No odds were ever too great to deter them from engaging the enemy who, day after day, threw himself with fury against the island. Daily they performed deeds of valour that will be talked about as long as courage and self-sacrifice count for any thing in the hearts of men.

New types of aircraft may come to retain the mastery of the skies won by the Hurricanes of the Middle East, but the pilots will never cease to pay their tribute to the old "Hurri-bus."

STILL IN SERVICE : A Westland Wapiti of the Indian Air Force, now adapted for target towing. Incidentally, the Indian Air Force celebrated its tenth birthday on the day of the R.A.F.'s jubilee.

Aircraft Types and Their Characteristics

HAWKER HURRICANE IID

NO aircraft has proved more adaptable to meet the increasingly wide range of duties which the progress of the war has demanded than has the Hawker Hurricane.

The latest version of the Hurricane about which information may be given is the IID, commonly known as the tank-buster, or still more affectionately as the tin-opener. This came into being specially to deal with the Hun's crawling arsenals in the African campaign (the biggest example of which was the Mark VI tank), and mounted a pair of 40 mm. cannon, one under each wing. The result of that campaign makes it unnecessary to stress how efficiently Jerry's tins were opened.

Basically, the Hurricane IID is the same Hurricane that began its belligerent career as a simple eight-gun fighter,

and has remained essentially the same through all its variations up to the present one. Possibly no better indication of its obliging nature could be given than its steady increase in all-up weight from less than 6,000 lb. to well over 8,200 lb. Actually, the weight of the tank-buster is rather less than the IIc with long-range tanks, but a little greater than the normal IIc.

Its 40 mm. cannon were designed specially for aircraft use, and at 640 lb. a pair of these weigh slightly more than four 20 mm. cannon. They are mounted in fairings beneath the wings, as shown below; and, since the IID was for service in Africa, the tropical air-cleaner was also fitted. Its top speed is in the region of 320 m.p.h. with the 1,280 h.p. Rolls-Royce Merlin XX engine, which is only a few m.p.h. slower than the IIc version.

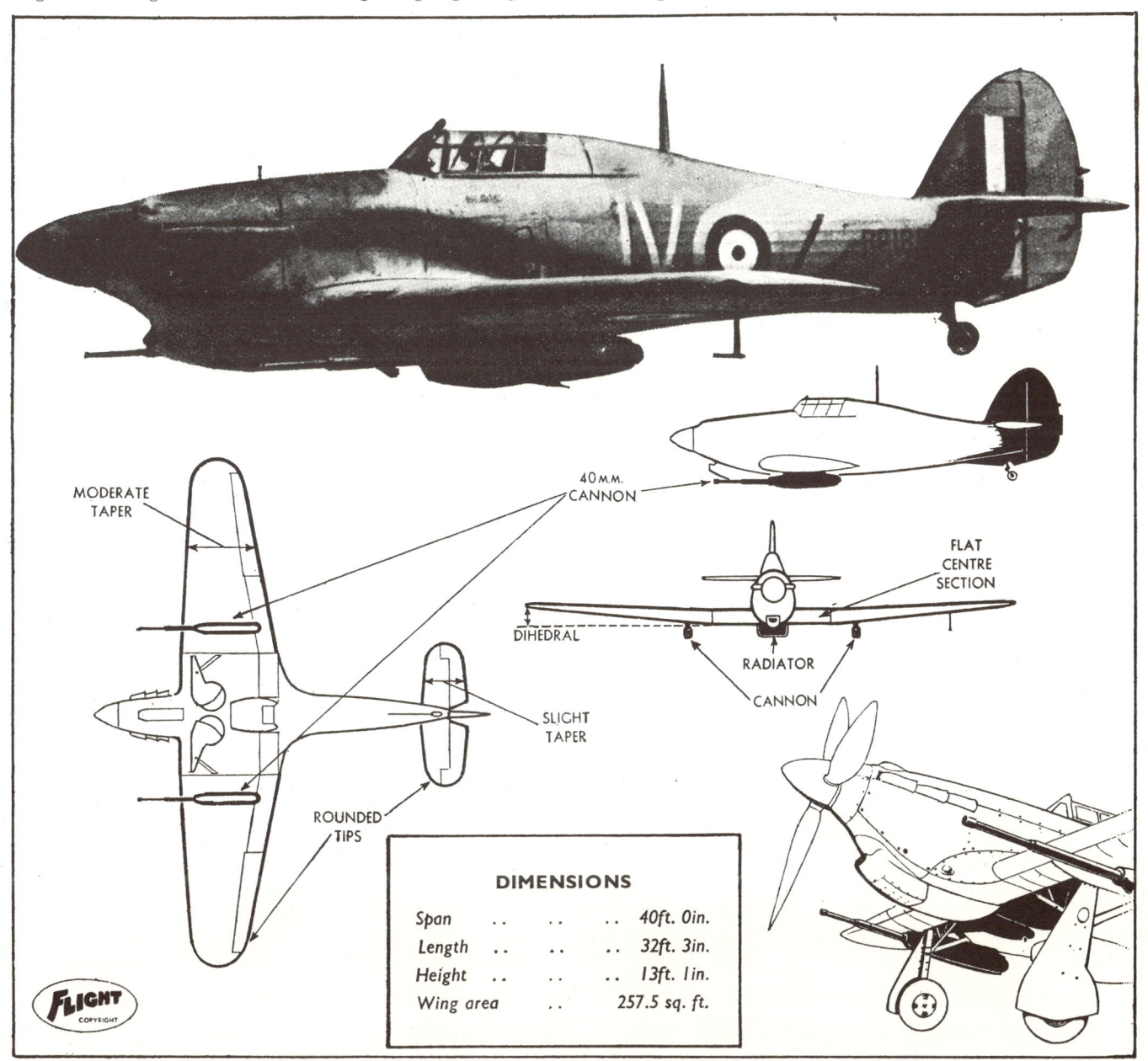

MODERATE TAPER

40 M.M. CANNON

FLAT CENTRE SECTION

DIHEDRAL

RADIATOR

CANNON

SLIGHT TAPER

ROUNDED TIPS

DIMENSIONS

Span	..	..	40ft. 0in.
Length	..	..	32ft. 3in.
Height	..	..	13ft. 1in.
Wing area		..	257.5 sq. ft.

FLIGHT COPYRIGHT

THE FIRST OF THE MANY: The prototype Hurricane, piloted by Flt. Lt. P. W. S. Bulman towards the end of 1935——

The Last of the Many

Hawker Factory Delivers Hurricane 10,000 Plus : Born 1935 and Still Going Strong

THIS week marks the delivery of the last Hurricane from the production line of Hawker Aircraft, Ltd. At long last this old warrior, which first flew in November, 1935, must give way to faster and more powerful single-seater fighters, but this does not mean that the Hurricane is to be pensioned off and retired from service. Still far from being relegated to the past, the Hurricane is even now a first-line aircraft in several of the many different roles it has undertaken. As a close-support, low-attack fighter it is daily in action with the land forces in Normandy, Italy and Burma ; as a fighter-bomber it carries two 500 lb. bombs ; and as a rocket-firing cannon-fighter its fire power is equal to that of the Typhoon, the effect of which has been compared to a broadside from a cruiser.

Of all the 24 different battle fronts on which the Hurricane has seen action, its greatest achievement came in the Battle of Britain in 1940, when Hurricane squadrons formed the main equipment of Fighter Command, and Hurricane pilots shot down more than half the total number of enemy aircraft destroyed. Without detracting in any way from the magnificent courage, skill and endurance of the fighter pilots, the British victory was in some measure due to Hawker's who produced the aircraft in time and in sufficient numbers to enable the pilots to turn the day.

To no less a degree, Hurricane pilots were responsible for beating back the onslaught of the combined *Luftwaffe* and *Regia Aeronautica* on Malta. When the position with regard to shipping losses was at its blackest, Hurricanes hastily fitted with catapult gear were mounted on merchant ships and were launched against enemy bombers

——AND THE LAST : No more Hurricanes will be turned out by the Hawker factory. The pilot is Group Capt. P. W. S. Bulman.

THE LAST OF THE MANY

which came out to raid the convoys. In the North African campaign Hurricanes played a leading part throughout, and towards the end the R.A.F. sprang a surprise with the Hurricane IID tank-buster armed with two 40-mm. high-velocity guns, which played havoc with enemy armour.

Of the total number of Hurricanes produced, which, by the way, is well over the 10,000 mark, a considerable proportion has been sent to the U.S.S.R., where the Russians, like ourselves, found that in the hands of skilful pilots the machine was more than a match for German fighters on account of its high degree of manœuvrability and powerful armament.

The first rocket projectiles to be fitted to a single-seater fighter were installed on a Hurricane, and on September 2nd, 1943, the first rocket attack by a squadron of Hurricanes was made on the dock gates of the Landweert Canal in Holland with results that were comparable to those of the Möhne Dam attack.

During the war, the Hurricane has proved itself in all theatres of action from the Arctic Circle to the Equator, from France to Burma and Ceylon, and back to France again. It has operated where no other fighter could go, from landing strips that have been little more than mud-patches, from snowfields, from desert sand. All credit is due to the ground crews who have kept them flying under

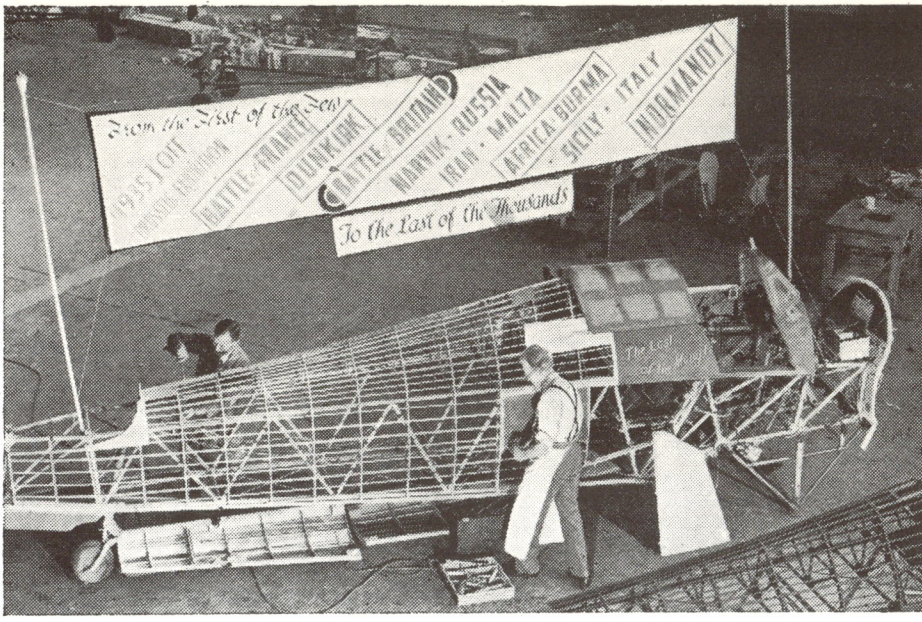

Hawker workers were somewhat sad at the thought of starting on their last Hurricane. The banner explains some of the reasons for their affection for the old war horse.

these conditions; they appeared to achieve the impossible, and if the machines had not been Hurricanes it would have been impossible.

The versatility of the Hurricane has never been surpassed by any other aircraft—one expert puts the number of variants at 163; although in reality this figure includes all the possible permutations and combinations of some twenty or thirty basically differing types. This versatility, combined with its high performance and fine flying qualities, together with the unyielding courage and tenacity of the fighter pilots, has made the Hurricane one of the most outstanding machines in the history of aviation.

preservation profile

No 40 Hawker Hurricane IIc

Preserved by the RAF

Above, brand new, Bulman in charge, 1944. Below, as G-AMAU, 1950.

As H3424/MI-G for Battle of Britain film, Hatfield, July 13, 1968.

As H3424/KV-M for Battle of Britain film, Duxford, August 30, 1968.

With the Battle of Britain Flight, Coltishall, 1972.

On September 15, 1944, when the Hawker Aircraft main assembly plant at Langley informed the RAF that Hurricane IIc PZ865 was ready for delivery into service, it marked the end of an era, for this was the last of 14,533 Hurricanes built. Powered by a 1,280 b.p. Rolls-Royce Merlin 20, it carried in its wings the four 20mm Oerlikon Cannon characteristic of this variant, and bore the legend 'The Last of the Many!' beneath the cockpit hood.

PZ865 was constructed in July 1944, and after being paraded through Kingston it made its maiden flight from Langley on July 27 that year, with Gp Capt P. W. S. Bulman, then Hawker Siddeley's chief test pilot, in the cockpit. After flight testing and acceptance by the Ministry of Aircraft Production, it was allotted to Hawker Aircraft, remaining at Langley for communication duties. In December 1945 it was purchased outright by the Company for private use, as had been agreed before its completion, then going into storage.

It was removed from storage in 1950 and entered by Princess Margaret for that year's King's Cup Race. Converted to meet ARB requirements as a civil aircraft, the cannon were removed, a whip aerial replaced the original mast, two extra 12½ gal fuel tanks were installed and the aircraft was given a smart royal blue and gold scheme with the civil registration G-AMAU. T. S. Wade flew it at the Royal Aeronautical Society's Garden Party at White Waltham on May 14, 1950, the day after its first flight in restored form, and it was issued with a C of A in the special category subdivisions (f) racing or record and (b) demonstration, on May 23.

On June 17, Gp Capt Peter Townsend flew G-AMAU to second place in the King's Cup Race at an average speed of 283 m.p.h., and later in the year it was given temporary production markings for its appearance, along with several Portuguese Hurricanes, in the film Angels One Five, about Hurricanes in the Battle of Britain. In August 1950 G-AMAU was re-engined with a Merlin 24 and flown to third place in the Kemsley Trophy Race at Swansea by Neville Duke, and in September it made the best actual time in the Daily Express Challenge Trophy Race.

It was raced on several occasions until 1953, and subsequently made numerous appearances at displays. Transferred to Dunsfold on November 21, 1956, by 1960 it was restored to its original wartime camouflage, but the cannon never reappeared and the civil registration was painted beneath the tailplane. Now flown as part of the "Hawker Museum" it was again re-engined in December 1962, this time with a Merlin 502. Disguised as H3424, and coded as MI-G of a Polish Squadron, and in sundry other markings, it took part in the filming of The Battle of Britain, shot at Duxford in 1968 and released on September 15, 1969.

On March 29, 1972, Hawker Siddeley presented PZ865 to the Royal Air Force Battle of Britain Flight, then based at Coltishall, and it became a stablemate of the Flight's Hurricane IIc LF363. It now bears the codes of 257 Squadron, based at North Weald during the Battle of Britain, and also carries the personal markings of Squadron Ldr Stanford Tuck, DSO, DFC, 257's CO at that time. Added to his own "kills" beneath the cockpit are two birdstrikes, one in 1972 and one last year.

Since September 1975 PZ865 has been grounded at Coltishall, pending the arrival of a new radiator, but it is hoped that it will be airworthy before the end of the current season. It will then be able to rejoin the Flight at its new Coningsby base, where it moved on March 1 this year. Although the four wing cannon are available, barrel fairings cannot be found to complete the installation, and they are still omitted.

50

Specification

Span	40ft 0in	Loaded weight	8,100lb
Length	32ft 0in	Maximum speed	336 m.p.h.
Empty weight	5,800lb	Service ceiling	35,600ft

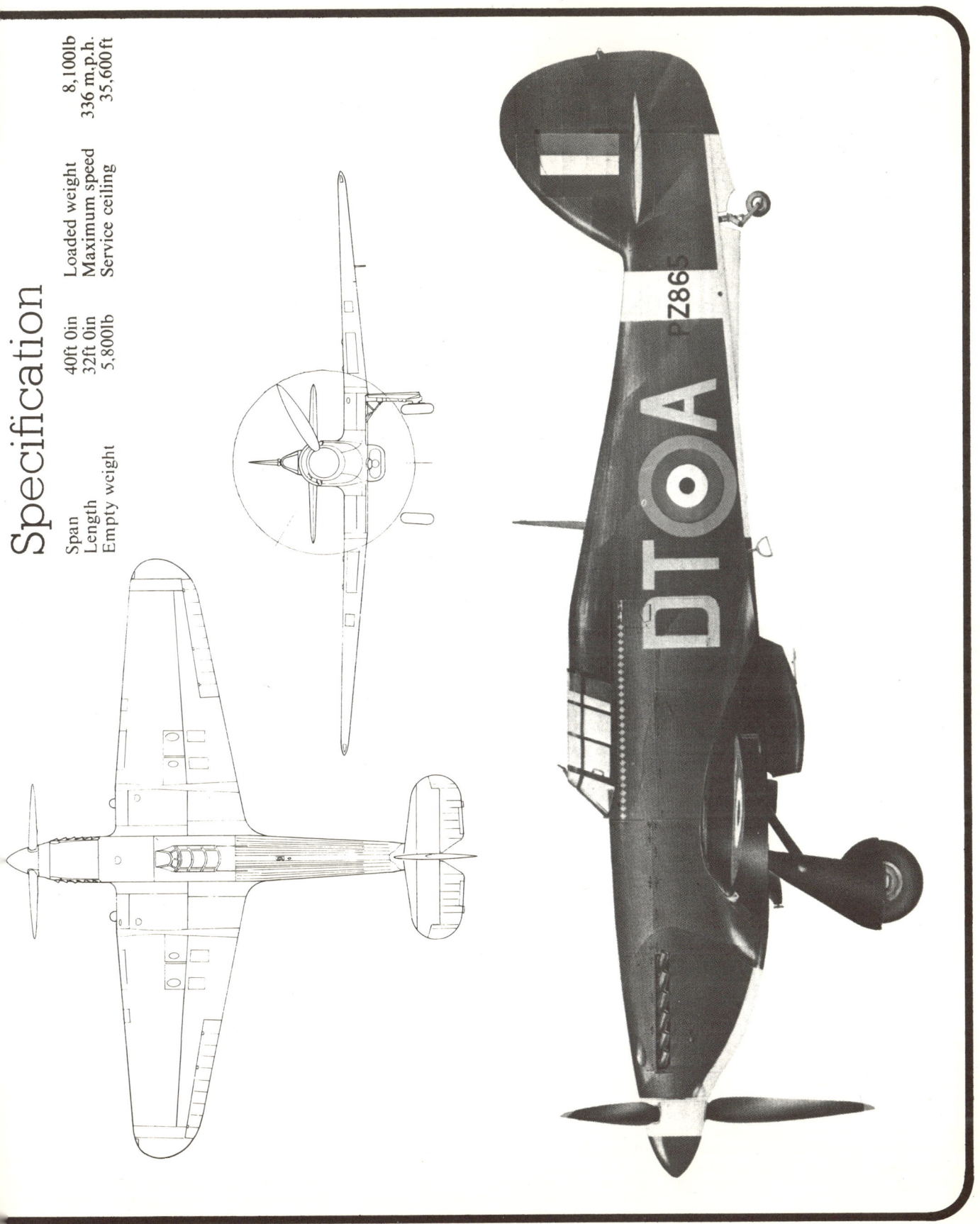

Field. We have been in touch with the Manager and have asked him to give you all the assistance that he can. Roberts Field is being built under the direction of a United States Army Construction Corps. You will have to remove all flashes and badges of rank as you must appear to be civilians. Liberia is neutral."

I left him and went straight to Hastings to see Sqn Ldr Billy Drake and arrange for suitable personnel to accompany me.

Flt Sgt Jones and Corporals Watson and Booker were selected. The next morning we boarded a DC-3 for Roberts Field with tool boxes, valises and all the impedimenta that we considered might come in handy.

That afternoon I went to see the Manager of the Firestone Rubber Plantation. He explained to me where the Hurricane was situated and arranged for a motor boat and crew to take us to the site, which was about 15 miles down the Great Bassa River. I told him that there were four of us and that I would like the boat to be available at eight o'clock the following morning complete with a picnic lunch.

Billets were arranged for us with the US Army contingent, which was under the control of two Sergeants, Overstone and Logsden. They could not have made us more welcome and were helpful in every way. It amused me because every morning at 0630 "Shake a Leg", "Rise and Shine", and every other kind of repartee burst upon us followed by a good tug, and as I raised my bleary-eyed head the voice invariably remarked, as if in complete astonishment, "Oh, sorry Sorr!"

The mess was like a large roadside café and the food was magnificent. A deep freeze which must have been 30ft long ran down one side and the tables ran down the other. Breakfast consisted of flapjacks covered with maple syrup, with bacon and eggs on top. Steaks were flown in specially from the United States. After the food in Freetown it was sublime.

THE HURRICANE AND THE SARDINE

A ir Headquarters West Africa was established in October 1941 in a two-storey house at Fourah Bay, Freetown. It was adjacent to Fourah Bay College which was occupied by No 95 (Sunderland) Squadron. As a Squadron Leader my exalted title was Command Engineer Officer. My staff consisted of a former grocer's assistant who typed with one finger. My office was an open balcony which I shared with Sqn Ldr Holmes, who was the Command Armaments Officer; we never got far with our paperwork because if it blew hard we spent the rest of the day retrieving the paper from the garden. A beautiful little green and orange Gambian parrot called William completed the complement of the combined Engineering and Armament Staff. In the rainy season I carried a large golfing umbrella and William accompanied me perched on the spokes.

About 13 miles east of Freetown was a landing strip called Hastings, occupied by a Fighter Wing equipped with Hurricanes

In May 1942 a Hawker Hurricane forced-landed in Liberia, West Africa. Its subsequent recovery is described in this article, taken from a manuscript preserved at the RAF Museum, Hendon; the photographs include some from an old roll of film which accompanied it. The manuscript carries no record of its author; it was selected from the uncatalogued section of the Museum's Department of Aviation Records by staff member SEB COX.

for the defence of the harbour.

One afternoon in May 1942 the AOC, Air Cdr Ed Rice, sent for me. "A Hurricane on its way from Takoradi to Freetown has made a forced landing about 15 miles south of Roberts Field in Liberia", he said. "I want you to go down and see if you can salve it. There is a large Firestone rubber plantation adjacent to Roberts

The morning after we arrived we went down to the jetty and awaiting us was a launch complete with everything that I had requested. It was about 25ft long and had a hard awning with open sides. The coxwain was a small negro dressed in an untidy looking vest and shorts, and wearing a sombrero type of hat. We set off down the Great Bassa River, which is about 150 yards wide. About 14 miles down river it branches to the left, but if you continue straight on you arrive at Monrovia. We took the left fork and after about three miles we came to a narrow cut in the reeds on the west bank; this was our destination. The river bank sloped upwards for two or three hundred yards to a native village. By means of signs and pidgin English we were able to explain that we wished someone to lead us to the Hurricane. We were at once surrounded by a number of willing volunteers. The ground was quite flat and a path led us through the bush to a large open clearing. There had been some rain and the Hurri-

Above, *the Hurricane up on its wheels again, apparently with the new propeller in place. Unfortunately the serial number is not visible, even on the original.*

cane was lying in a corner on its belly with the lower part of the engine cowling in a pool of water. The tips of all three propeller blades were broken.

I had earned my living for the six years prior to the war as a chartor pilot. Getting in and out of fields and off race courses was nothing new. It had been my lot to fly many and varied types of aircraft. Luckily Billy Drake had allowed me on one or two occasions to fly one of his Hurricanes, so I was conversant with the "taps".

It occurred to me at once that if the clearing had a reasonat'e surface, and if the engine could be made serviceable, there could be a possibility of flying the aeroplane out. Leaving Flt Sgt Jones and his companions to assess the damage, I set of across the clearing to find out whether my idea was practicable. In the main the surface was sand—quite flat but with some soft patches where the water came over the top of my shoes. There appeared to be a straight run of about 800 to 1,000 yards free of obstruction. I decided that if there was no rain before the attempt was

The locals are "apprenticed" to help lift the new propeller from its makeshift assembly bench, **below**. *The condenser is* **below right**, *with some of the lettering on the sardine-tin still perfectly legible. It is preserved at the RAF Museum.*

made, then the aircraft could be flown out. Flt Sgt Jones confirmed that, as far as he could see, there was no serious damage apart from the propeller. However, it was essential to get the aircraft jacked up and the undercarriage down before any final opinion could be arrived at. We returned to Roberts Field full of hope, and prayed that it wouldn't rain.

One thing was certain: if we were to fly the aircraft out, we had to obtain a new propeller; so a telegram was sent to Takoradi requesting the Maintenance Unit there to send one to Roberts Field.

Jacking the aircraft up was going to be difficult. It would mean digging a trench under the wing to get in a small car or lorry jack and raising the machine inches at a time until a larger jack could be used. I found two invaluable pieces of equipment in the form of two empty wooden cable drums and these, together with an assortment of pieces of wood, formed the basis of our equipment. The cable drums could be floated ashore and rolled to the site. They were used as platforms on which to complete the jacking-up procedure, and subsequently made a most useful workbench on which to assemble the new propeller.

At that time Pan American Airways and Trans-World Airlines had servicing personnel at Roberts Field. Red McKenny was a Trans-World employee and as TW flights were somewhat infrequent he offered to give us his assistance—which

we were delighted to accept. He was not only a first class mechanic but the possessor of a toolkit *par excellence,* way in advance of anything that was issued to RAF groundcrew. He and his toolkit saved us many hours' work on the job before us.

I heard in the course of conversation in the mess that there was a German storekeeper in Monrovia who had built himself a lookout tower and was reporting shipping movements to his military compatriots. I decided, on my way to the Hurricane site, to pay him a visit. It was easy to find the store because, sure enough, there alongside it was structure about 30ft high commanding a view out over the sea. Cape Palmas, some way to the south, was at the limit of the range of our Sunderlands, and a number of ships had been sunk in the area.

I went into the store and passed the time of day with the owner, bought something trivial and left. When I returned to Freetown I made a full report on my visit and I am glad to say that as a result of pressure from the Americans it was not long before the Liberians closed him down.

As we worked on the Hurricane large cumulus clouds collected and things looked ominous, but the rain held off. Within two or three days the aircraft was standing on its wheels with the undercarriage down and locked. The propeller had arrived and our native friends from the village carried it in its dismantled state on their heads to the site. The radiator was full of sand and we sat underneath with bamboo slivers and poked it out until the matrix was clear. Everything was checked and appeared to be in order. There was very little petrol in the tanks. We half filled two drums at Roberts Field so that they could be floated ashore and rolled to the site. Talk about "Saunders of the River"! You could hear our friends rolling the barrels through the bush for miles. Every evening Jones and I had a pay parade. Those carrying light loads got 3d, and the heavy load carriers got 6d. We paid them in English money; I have no idea of what use they made of it—possibly earrings! In any case everyone appeared to be satisfied.

Guns, ammunition and everything else

Cowlings being removed out in the midday sun, **above**; *note the truncated propeller blade. The evidently blistering heat probably accounts for the deterioration of the original roll of film, and hence for the poor reproduction.*

that was weighty and could be removed was removed. At last the great moment arrived to start the engine. It wouldn't. The ignition was checked and the spark was found to be very weak. It was lunch-time and we sat down to eat and to puzzle over the problem. Our rations included some tinned sardines and some chocolate wrapped in waxed paper. After some discussion it was the general view that, as the booster coil was situated at the bottom of the engine bay, its condenser had probably been lying in water and hence was unserviceable.

It was not the sort of spare part that was readily available and, besides it might take several days to get a replacement; each day we could expect rain, so we could not afford to wait. Red McKenny suggested that we should make a condenser out of the sardine cans using the waxed chocolate paper as a dielectric. We all sat round cutting the cans into strips. Red assembled the finished product, wrapped it round with some insulation tape and, after Jones had attached it to the booster coil, all was ready for another try to start the engine. Jones got into the cockpit, settled himself down, pressed the starter button and choof, choof—two compressions and a cloud of smoke, and there we stood watching the engine running as sweet as a nut. When he had given the engine time to warm up Jones opened the throttle and the engine roared away; but all was not well. The clutch driving the low altitude supercharger was slipping, and it was impossible to get any boost. Jones next discovered that, by using the high altitude clutch, normal take-off revs could be achieved at maximum permitted boost. We ran the engine for some time and apart from the clutch trouble all seemed to satisfactory. I decided that we would wire up the over-riding boost controls so that if I looked like hitting a tree on take-off I could momentarily over-boost the engine with a good chance of getting away with it. There was, of course, very little throttle movement required, and a careful eye had to be kept on the boost gauge needle.

Some pilots believed that if, after turning into wind, they put on their brakes and opened the throttle fully, and then released the brakes, this would result in the shortest possible take-off. I never subscribed to this view. My method had always been to taxy as fast as possible, to make a turn into wind and then open the throttle fully, without braking. The propeller was then never running fully stalled. Using my usual technique I set off to the downwind end of the clearing. The tail lifted as I went over a soft patch and I said to myself, "You will have to watch out for that. It would be very ignominious to end up on your nose." Round into wind at the end of the clearing; throttle open; check boost gauge: 7.5lb; revs normal; away we go. Another soft patch! Tail lifts, hold it ... that's good, starting to lift; plenty of room; over the trees and away. Shall I retract the undercarriage? No, better wait until a thorough check has been made. The Great Bassa River was there below me. Roberts Field soon appeared and so down onto the runway, rather pleased with myself.

As soon as the others arrived by boat the story of the sardine can condenser made the rounds and a splendid evening of ballyhoo ensued. Red McKenny was the magician of the hour and the pair of us enjoyed free drinks and much leg-pulling until late into the night.

The next morning, leaving Jones and the rest of the party to change the engine for a new one to be sent from Takoradi, I returned to Air Headquarters. "So you managed to salvage the Hurricane," said the AOC as I entered his office.

"Yes Sir"; I told him the story of the sardine can condenser and the tremendous help that McKenny had been to us. "Good", he said; I saluted, and returned to my veranda to write a letter of appreciation to TWA and to see William the parrot.

STUART HOWE'S **colour plate** *and* **heading photograph** *are the most recent photographs of the CWH Hurricane and show the aircraft in its new markings. Aircraft P3069, coded YO-A, represents a Hurricane of No 1 (City of Westmount) Squadron.*

HURRICANE RETURNS HOME

The latest acquisition of the Hamilton Airport-based Canadian Warplane Heritage is the ex-Strathallan Hawker Hurricane IIB "P3308". This particular aircraft is no stranger to Canada as it was one of 1,451 Hurricanes licence-built by the Canadian Car & Foundry Ltd at Fort William from 1940 onwards.

Powered by a Packard-built Merlin 29, Hurricane 5377, construction number 42012, was first delivered to No 2 Training Command and put into storage until the following year, when it was put into service with 163 Squadron. With this unit it flew coastal defence and Army Co-operation sorties at Sea Island, British Columbia, during which time the Hurricane was re-designated Mk XII.

After the war many Hurricanes became surplus to requirements and a large number became dispersed on farms around Canada where they became a useful source of spares for farm vehicles and equipment. Hurricane 5588 spent some time at Portage La Prairie where it was held in reserve for display purposes. In 1952, with less than 100hr total flying time, the Hurricane was sold to Ajax Aircraft Parts and was eventually bought by a farmer.

In 1964 the Hurricane was discovered by aircraft collector and restorer Bob Diemert. He purchased the Hurricane and then set about restoring it to airworthy condition. A 1,634 h.p. Merlin 25 was installed and the Canadian civil registration CF-SMI applied. After a short period flying in Canada the

Hurricane was sold in the UK for use in the film *The Battle of Britain*, arriving in this country in 1967. After filming had been completed the Hurricane was sold to "Tony" Samuelson and hangared at Elstree in company with his pair of two-seat Spitfires G-AVAV and G-AWGB. Registered G-AWLW, the Hurricane had a certificate of airworthiness issued to it in May 1969. In December the Hurricane was purchased by Sir William Roberts as part of a package deal which included the aforementioned two-seat Spitfires. These three aircraft were to form the nucleus of the Strathallan Collection and they were initially based at Shore-

ham in Sussex. In March 1972 the Hurricane was transported to Strathallan by road for restoration. This was completed in about a year and on June 28, 1973 test pilot Duncan Simpson took the Hurricane into the air once more. Painted up to represent Hurricane P3308, coded UP-A, the aircraft flown by 605 Squadron pilot Sqn Ldr Archie McKellan, the Hurricane remained with the collection until it was put up for sale in 1984.

It was sold to the Canadian Warplane Heritage for C$450,000 and in May 1984 returned to Canada aboard a Hercules transport aircraft.

In Canada the Hurricane was re-registered CF-SMI and painted to represent an aircraft of No 1 (*City of Westmount*) Squadron, P3069, coded YO-A. This unit was the sole Canadian squadron to serve in the Battle of Britain. The Hurricane took to the Canadian skies once more on June 4, 1984 and, in the hands of Rick Franks, will no doubt delight Canadian airshow crowds for years to come.

A. Denholm

Dick Richardson

Right, *two photographs of the CWH Hurricane during Strathallan ownership. The take-off picture was taken during an air display at Strathallan on May 27, 1979.*

Y
A
P3069

Y
A
HAMILTON F
PIPER

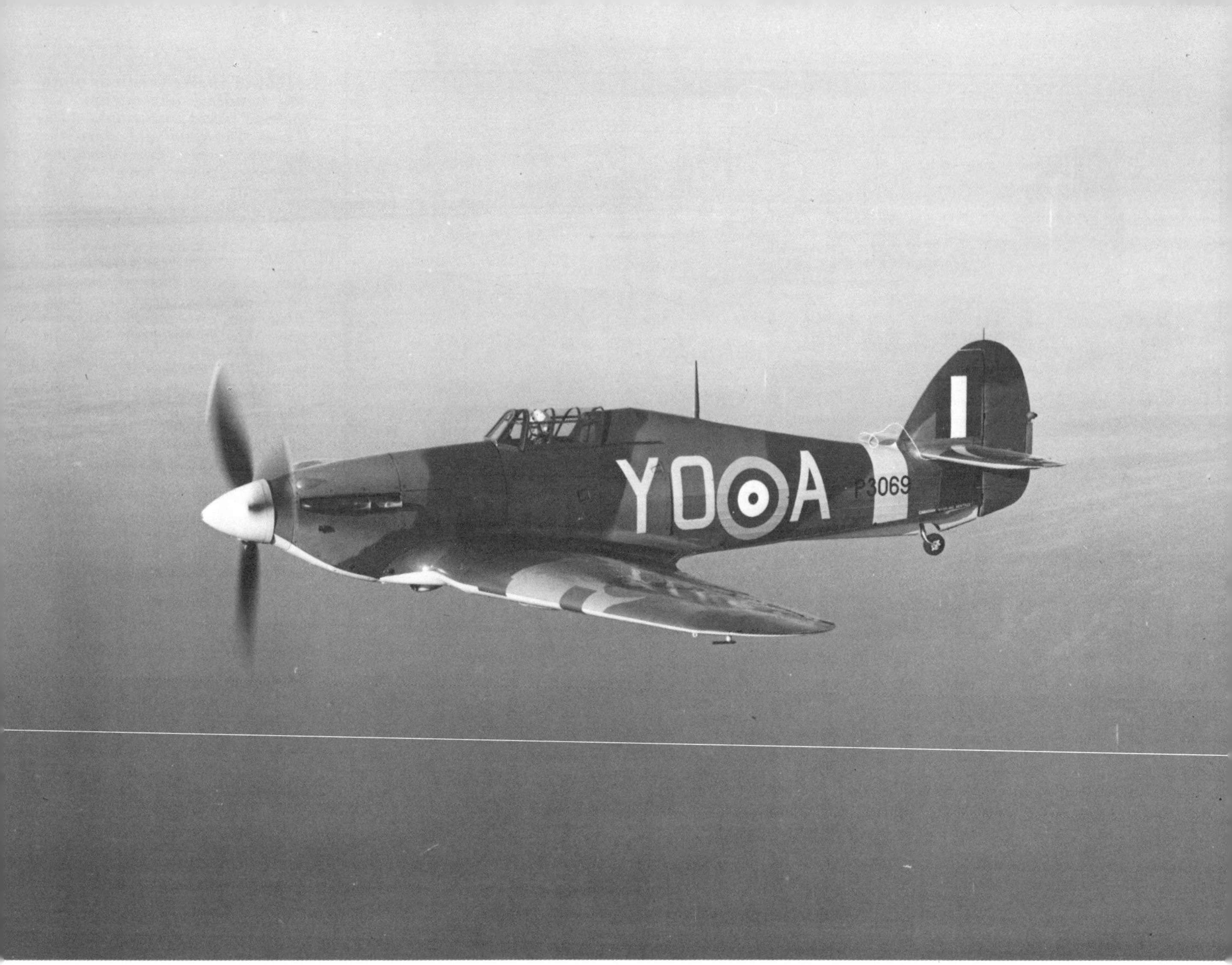

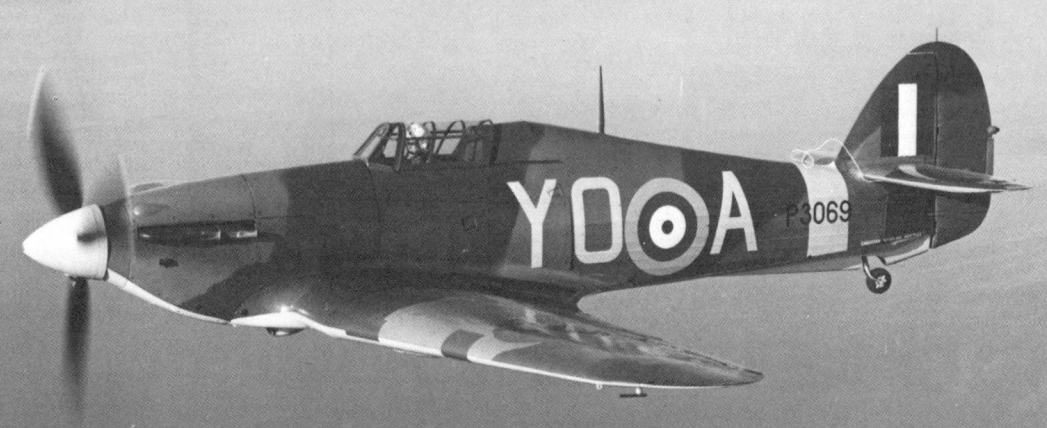

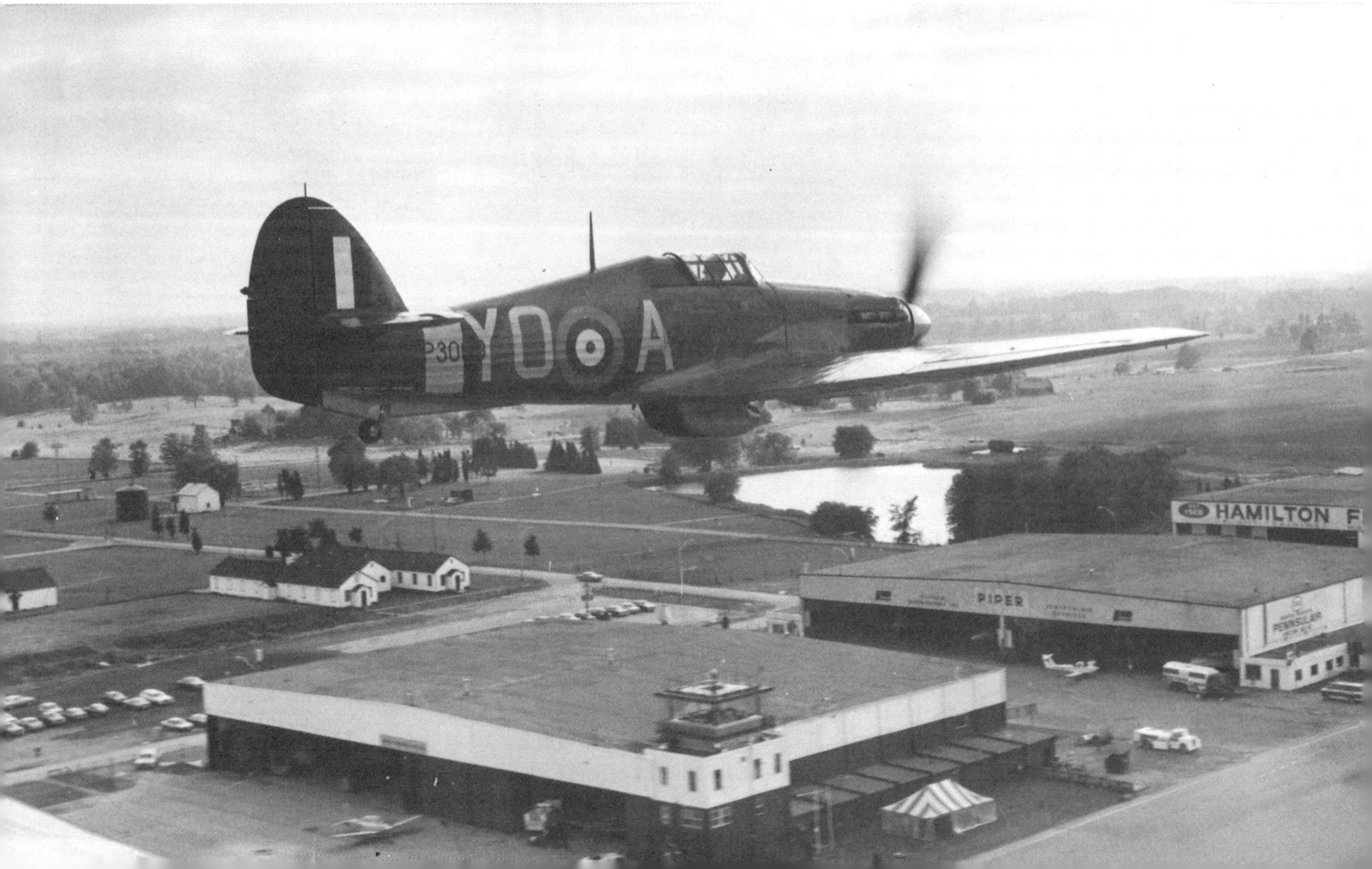

HURRICANES AT NIGHT

WG CDR ROLAND BEAMONT begins a two-part account of 87 Squadron's Hurricane night fighter operations from RAF Exeter and the Scilly Isles during 1940–41

The onset of winter in 1940 was a strange period for 87 Sqn. We had spent the previous six months, firstly in the thick of the day fighting in Northern France and Belgium and then, after a short pause for breath, beer and re-equipment in the tranquillity of the Yorkshire countryside at Church Fenton, back into what was quickly to develop into the right flank of the Battle of Britain, from Exeter.

Losses in France in May and then through July, August and September in major battles over the Channel from Portland to Portsmouth had totalled more than 50 per cent, including two fine commanding officers, "Johnny" Dewar and Terry Lovell Grieg. But we gave as good as we got or better and, by October, with the emerging realisation that the Luftwaffe had been defeated, the morale which had been inspiring and unfaltering throughout the summer's battles was sky-high indeed.

At that point movement orders were received to base the squadron at Charmy Down, a new satellite airfield of RAF Colerne near Bath, with some intermediate operations from the grass airfield at Bibury whence we had carried out periodic moonlight patrols over Bristol during the past few weeks. This looked ominous, and it was soon confirmed that 87 was to become a full-time night-fighter squadron!

This prospect seemed disastrous to a bunch of day-fighter pilots who had not only survived the two great air defence actions of the war so far, but had done so with a very positive share in the ultimate victory. As they now saw it, they had to give up the inspiring climbs in formation into the brilliant skies above cloud to fight and throw back an arrogant enemy whose black-swastika'd hordes had daily trespassed over our land and homes. Instead they would have to live a life of crouching in the dark in ill-heated huts, awaiting the telephone call from Ops to "Patrol Line A at 10,000ft"—an order which would lead to a lone flight into the darkness in whatever the weather happened to be when the order was issued. No 10 Group Headquarters' comforting observation in this connection was already well known: "If the Hun is flying, so can you!".

It was an enormous contrast, and it said much for RAF training, as night bombing was seen to be the new threat to this country following the failure of the German air forces against Fighter Command in daytime, that 87 and the other squadrons concerned just got down to the new job as best they could.

But all was not well. There were no effective homing aids at the beginning and each sortie had to be conducted by DR (watch and compass) navigation to and from the patrol line which, in theory, could be seen from 10,000ft as groups of flares at 10 mile intervals.

In practice, of course, the flares could only be seen on very clear nights—which seldom happen over this country in the winter. So the patrols were flown by timed runs on reciprocal headings and, in the cases of all but the most skilled in pilot navigation, after $1\frac{1}{4}$hr on patrol on a pitch dark night and over cloud or mist there were few pilots who had much of an idea of their position.

In theory a "fix" should have been possible by triangulation on radio voice transmissions, but in practice the TR.9 radios were so unreliable and sensitive to atmospheric conditions that the night fighter pilot would have to resort to setting the "safety course" for his home beacon at the end of his $1\frac{1}{2}$hr patrol and, beginning his descent on a timed run, then hope to break out below cloud in sight of the

Richt, *Fg Off "Watty" Watson with his 87 Sqn Hurricane at Exeter in 1940. Note the absence of the rear-view mirror, removed from the top of the windscreen to reduce drag.*

beacon flashing his base signal—or any other beacon which would lead him to somewhere to land before his fuel ran out!

Once on patrol, of course, it was a case of eyeballing with no radar through a thick armour-glass windscreen surrounded by heavy metal structure and the reflector gunsight (with bright aiming spot graticules when switched on). To sight and hold on to another aircraft was supremely difficult. The majority of our few interceptions in that winter were in fact on searchlight-illuminated targets or following fleeting glimpses of bombers silhouetted against the massive fires of the cities of Bristol, Swansea, Plymouth or London burning below.

Intense effort

Nevertheless, the job was tackled with intense effort and what would nowadays be called "professionalism", and specific training was carried out on every suitable night by pilots not on standby. This included local and cross-country formation flying in moonlight and full darkness and, of particular importance, pairs flying to assess each other's limits of night visibility. The No 2 would drop slowly astern of the leader until the latter's silhouette disappeared at about 400yd, and then the blue exhaust flames became almost invisible. Inevitably contact was sometimes lost, and in all these sorties DR navigation over a totally blacked out countryside was the sole means of recovery to a safe landing.

The enthralling vistas of the day fighter pilot in wide open skies had been replaced for us by the confines of the dimly red-lit cockpit with often nothing distinguishable in the total darkness outside. Our whole existence seemed limited to this cramped space and to the thunderous vibrations of the Merlin engine on whose continued

Above, *a 213 Sqn Hurricane which crashed into B Flight dispersal of 87 Sqn at Exeter in August 1940, on return from combat over Portland, Dorset.*

roar life indeed depended—for, in our experience, parachute escape from the Hurricane was by no means an assured way out.

Clear moonlight, or even very clear starlight nights were a pleasure, as navigation could generally be visual on the outlines of the Severn, or on occasions Plymouth Sound or the Thames. However, in all other circumstances of darkness in mist, rain, thick cloud, icing and snow, all of which occurred in plenty in that winter at Charmy Down, uncertainty was dominant until, sometimes, after casting around in a black goldfish bowl at the lowest altimeter height considered safe relative to the hills one might be over, the mist to one side or ahead began to glow intermittently with the signal of a flashing beacon. Then it was turn up the cockpit rheostat, check the beacon signal against the beacon card for the bearing to the airfield, set the new course and begin final descent. Then, below 500ft, the line of shrouded paraffin flares appeared one after the other ahead. Undercarriage and flaps down; sometimes a hurried S turn to line up; throttle closed, stick back, sparks each side from the throttled engine, then the thump of a hard

Left, *Sqn Ldr Ian Gleed with A Flight pilots at Bibury in October 1940. Left to right, Watson, Tait, Gleed, Rayner and Comely.*

arrival. Less frequently, touchdown was a gentle three-pointer and there was no more to do but try to keep straight between the flares—not all that easy in a crosswind in the wet.

After landing there would be a brief handover to the loyal groundcrew, who would service one aircraft after another all through the night in the open, whatever the fierce weather. Then it was back to the generally bitter-cold dispersal hut for a hot cup of tea and a lie down with a blanket for the rest of the night—or until one's turn for another patrol, this time, with luck, to return in the light of the dawn.

It was a challenge every night, but although the risk of accident was high and many aircraft and some pilots were lost that winter, the chances of operational success were low. Of the very few interceptions which occurred, most were inconclusive. The writer fired at just two Ju 88s in the searchlights over Bristol during a five month period; both plots disappeared over the Severn and were recorded as "probably destroyed". But when "Splinters" Smallwood saw some navigation lights over the middle of a major raid on Bristol one night and dived to investigate, he found "a clot of a Heinkel" and shot it down over land.

It was slow going, and by the spring of 1941, with the news that the Spitfire wings were taking the offensive and beginning "the sweeping season" over France, 87 sought other outlets from the dreary round of night "Politician Patrols".

The energetic CO, "Widge" Gleed, obtained tentative and rather surprised support from 10 Group Headquarters at Rudloe Manor for two experimental operations, the future of either of which was to depend "strictly on results"—which meant that in the event of initial failure there would be no second chances.

The first plan was to send a detachment to a small grass airfield on St Mary's in the Scilly Isles, for the purpose of day interception of enemy mine-laying and reconnaissance sorties which had been reported in the area in some numbers by Intelligence recently. The second plan was for that time an unusual and, some thought, extremely radical one of employing our new-found confidence in night operations in ground attacks by moonlight on enemy airfields.

Both these interesting operations began in May 1941 and are the subject of Part Two of this article in the next issue.

59

By the spring of 1941, following a long, hard winter of night operations and a training syllabus which had even included regular night formation aerobatics, 87 Sqn felt ready to extend its newfound expertise. On April 9 the first of a new type of operation was mounted.

Sqn Ldr Ian Gleed led four aircraft to Warmwell at dusk for refuelling, and then there was a long wait until the take-off time planned for the first sortie. The plan involved the CO and his usual No 2, "Rubber" Thoroughgood, reconnoitring the area of Caen-Carpiquet airfield south-east of the Cherbourg Peninsula, and attacking targets of opportunity if the nearly-full moon gave sufficient visibility. If the defences were active and included searchlights, each aircraft would attempt to cover the other by strafing.

No-one knew whether ground-attack with 0·303in machine guns would be practicable by moonlight. The object was to

direction. Turning sharply, he had seen exhaust flames against the stars; he closed in and opened fire on a clearly recognised Dornier Do 17.

Thoroughgood meanwhile had lost his leader in this manoeuvre, but then saw a fire going down and hitting the sea, so he also came back.

Gleed, having had one successful sortie and now encountering a delay in re-arming, was of a mind to try the offensive operation on another night, but New Zealander Derek Ward was not to be put off and it was decided that the second sortie would go as planned—so the second section would now do the exploratory work!

With a now brilliant moon and a stable weather forecast all was favourable and, after a loose formation take-off over the undulating Warmwell grass and setting course on the climb for France, station-keeping in the moonlight was as easy as in daylight.

coast, and time to tighten the harness straps, check engine and fuel gauges, switch on gunsight with rheostat set low hopefully for ground-attack, and then finally turn the gun button on the spade grip from "Safe" to "Fire".

With the coastline clear below, Ward's Hurricane began to lose height gently, holding course for Carpiquet which should come up in 4min. Down to 2,000ft, then with final confirmation of target area two brilliant blue searchlights snapped on and weaved almost horizontally ahead in agitated scan.

Breaking radio silence Ward called, "There's the aerodrome", and dived down to port. Ahead the flak defences erupted in chains of tracer weaving haphazardly at first, and then one of the searchlights illuminated the leading Hurricane followed immediately by another and it looked for all the world like a moth twisting in a car's headlights.

Here was the No 2's task, and rolling

HURRICANES AT NIGHT

find out, and if the CO's sortie was successful Derek Ward and the writer would make a follow-up attack. This was to be not only the squadron's first night ground attack, but probably the first by any single-engined fighters in World War Two.

At Warmwell, in the stillness of the otherwise deserted aerodrome, the moon rose in a cloudless sky over the Purbeck Hills and the chill of a heavy dew heralded a frost later, leading to practical thoughts of keeping windscreen and wings protected. Then start-up time approached, and the first two Merlins crackled into life, shattering the stillness. Presently they taxied out over the rough grass, turned into wind and thundered off in loose formation, extinguishing navigation lights as they turned south under the now bright moon.

Ward and the writer prepared for a long 2hr wait for the first pair to return, and for confirmation that the next sortie would be on; but as we were entering the cold Mess building, otherwise deserted on this day-fighter station, the unmistakable sound of a Merlin approaching brought us back to the "flights" in a hurry and in time to see LK-A, Gleed's Hurricane, swing round and switch off.

He slid back the hood as we climbed up on the wing and said, "Got a bloody Dornier off Lulworth! Where's Rubber?". But then the other Hurricane curved in down the moon-path, and in a hilarious debriefing it transpired that hardly had they settled down on course for Normandy at about 10,000ft than Gleed had glimpsed a dark shape going past in the opposite

RICHARD WINSLADE'S colour plate, *taken on June 28, depicts Hurricane IIc LF363 of the Battle of Britain Memorial Flight. The aircraft has worn the black night-fighter colour scheme of 85 Sqn for the last three seasons—compare this picture with the colour plate on page 409 of last month's issue.*

WG CDR ROLAND BEAMONT concludes his two-part account of 87 Squadron's Hurricane night fighter operations from RAF Exeter and the Scilly Isles in 1940-41

With 80 miles of water ahead the customary illusion of engine roughness soon occurred, to be disciplined after a close scrutiny of the perfectly healthy engine instruments.

Cross-country night formation carried no mystery for us and the writer, flying No 2, relaxed in the comfortable knowledge that the leader's navigation would be accurate. It was sufficient to cross-check only at the first visual check point on crossing the enemy coast to ensure a good "safety course" for the likely solo return, because visual contact would almost certainly be lost over the target.

The two Hurricanes droned steadily on at 10,000ft and then, indistinctly at first and soon with final definition, a thin dark line to starboard revealed the Cherbourg coast curving round eastwards towards the Normandy beaches. It was the enemy

down onto the source of the nearest searchlight the writer aimed directly at it and fired a long burst with first the tracer and then the de Wilde flashing explosive rounds confirming accuracy. With breath-taking suddenness the light snapped out leaving a dying glow which helped judgment of pull-up; and then snap—the other light swung right on from almost dead ahead. In the dazzling glare nothing could be read in the cockpit, and, though at very low altitude, the ground was invisible. Tracer shells now began to whip by with a pronounced "whoomph, whoomph", and the only possible action was to aim and fire straight down the beam, knowing that this was the classically dangerous manoeuvre, as it would destroy night vision for a probably fatal period if and when the light went out.

After a seemingly endless burst, but in reality a very few seconds, the dazzle snapped out to blackness, again with the dying glow of the light helping to avoid flying right into it; and then with a snatched pull on the stick the writer had fleeting impression of objects flashing by on each side, before regaining some sort of re-orientation in a shallow climb with the moon in the right place and the altimeter now perceived to read 300ft and climbing.

But where was the target? Over to port, converging chains of tracer shells showed where Ward might be and immediately lines of tracer pouring downwards showed his attack on something—then a flash of fire on the ground. There were no searchlights now and the writer, quickly aiming for the fire and diving down through 200ft, suddenly saw the moonlight glinting on runways, hangars and the shapes of parked aircraft by the now raging fire.

With aiming-spot near the fire a continuous gun-burst strafing run was made through what looked like parked Me 109s and on into a hangar, before clearing low over the latter as the Brownings stuttered into silence, out of ammunition.

Keeping the Hurricane low, and pursued briefly by chains of tracer shells and bullets, the writer flew over shadowy fields, woods, a railway line with red-lit signals, a glinting canal and with final relief the light strip of beach at the coast. He then pulled up at climb power to set course for the planned recovery base, Middle Wallop, with only a single searchlight still weaving behind.

On the climb a disturbing hot smell caused swift study of the engine panel and outside for any signs of battle damage, although no hits had been felt; but all was well and the smell was recognised as cordite smoke from the Brownings.

The outbound crossing had been on planned time, and now the navigation had to be good to find the new landing base,

"Operation Fishing". *No 87 Squadron's readiness section at lunch on the Scillies in May 1941, on the cliffs immediately below the three Hurricanes parked on the edge of St Mary's aerodrome. Left to right, Roddy Rayner, author, Watty Watson and John Strachey, squadron adjutant and later a minister in the post-war Labour government.*

Middle Wallop. A brief radio call to Ward produced only silence and worry—his aircraft should not be far away, so had he been hit?

Under the still clear moon, but in haze up to 8,000ft, the sky and sea below seemed to merge and only in a look back did the moon-path on the sea confirm that there was no cloud below.

Half an hour of holding course at 10,000ft seemed ages longer, and then in a gentle descent the greyness ahead was searched for a sign of land. A recognisable pin-point was important for confirmation of heading to the Middle Wallop beacon.

Suddenly a faint white line appeared to starboard. This seemed wrong, as it should have been the chalk cliffs of Purbeck at Swanage's Old Harry Rocks to port. This must be the Isle of Wight at the Needles, and descent rate was increased to ensure a better view of the mainland while crossing in, but in the haze this could not be seen.

Nothing to do but hold course until within radio range of Wallop and then suddenly—duck! A large black shape loomed ahead and rushed down the starboard side. Then another straight ahead and a third above to port.

In the concentration on events the Southampton balloon barrage had been completely forgotten! In clear weather they flew to above 5,000ft and a glance at the altimeter showed 4,500ft!

With full throttle and fine pitch the

Right, the author's Hurricane at Bibury, October 1940. Note the exhaust flame shields for night flying, the lack of windscreen mirror and the combat score (5) recorded below the cockpit-sill.

Hurricane was stood on its tail and, with flashing glimpses of more balloons, this time their silvered tops seen from level and above glinting in the moon, the danger was past.

Ahead was a flashing beacon reading the code for Wallop, and soon the Hurricane was bumping down the strange grass flarepath. The time was 0200hr and at dispersal (John Cunningham's Beaufighters) no news was to be had of Ward. While LK-L was being refuelled and re-armed, experiences were exchanged with the Airborne Interception-equipped Beaufighter pilots who were intrigued to hear about our attempts at "eyeballing" enemy aircraft at night and the night ground attacks in the single-engined Hurricane. They thought we were mad on both counts.

Then there was a short, relaxed, 15-minute flight back to Bath under the still bright moon to land on the Charmy Down flarepath before dawn, and to find Derek Ward already there after returning via Warmwell where he had landed to check for possible flak damage.

He said, "Thanks for shooting that searchlight off me Bee—it was great. We must try that again!". As indeed we did, on May 7, again from Warmwell, when the writer led an attack on the Me 109 airfield at Mapertus on Cherbourg in LK-L (serial V7285) with Pete Roscoe flying No 2.

There was no difficulty in seeing runways and buildings in the moonlight but, before we could identify aircraft or other targets, searchlights and flak were very active and accurate. Both aircraft fired briefly at gunposts and searchlights before disengaging discreetly.

Shortly after crossing out over the coast a brief winking signal-light revealed the white V-shaped wake of a fast vessel travelling west close in to the shore. On the assumption that it might be an E-boat, the writer eased into a diving turn to line up the target against the moon path. This worked out and, at a few hundred feet up and an estimated 1,000yd away, the low profile of a fast patrol boat could be easily seen in silhouette. Fire was opened and continued to point blank range, with de Wilde hits sparking all round the bridge area and tracer ricocheting upwards.

The ammunition ran out as the Hurricane cleared close overhead and was held low in a tight left bank away from the moon as late return tracer fire erupted from one or two gun positions. But soon these fell away behind, and course was set for Warmwell which was reached uneventfully.

The most hazardous part of the operation, however, occurred at Warmwell after refuelling and re-arming, when during the take-off over white frosted grass towards the misty Purbeck hills the right wing dropped violently. Heavy vibration set in, accompanied by a rumbling roar, and nearly full left stick was suddenly required to hold the wings level.

Even in the indistinct light of the moon the cause of the trouble was evident—a large dark hole in the top surface of the starboard wing showed where the whole gun servicing panel on that side had disappeared. So the problem was twofold: how to make a safe landing, and where.

Some experimenting showed that with undercarriage down and no flap, aileron control of the wing-drop ran out below 100 m.p.h., and this would mean a fast landing although flap could be lowered once the wheels were on the ground. It was therefore decided that the longer runway of Charmy Down would be better for this sort of arrival, and also it would be less complicated to do it at home!

So course was set for Bath, in gathering moon haze which was not helpful, and, after some anxious moments, the home beacon was eventually seen periodically illuminating the murk ahead. A long, flat approach was made towards the Charmy Down flarepath, which thankfully appeared out of the haze about a mile ahead.

At 110 m.p.h. the left stick load was becoming nearly unbearable and when the threshold lights flashed by underneath the flaps were selected as the main wheels touched in a tail-high "wheeler".

With the red "glim lamps" of the runway-end approaching rapidly, maximum possible braking was applied short of nosing over, and then, with hard left rudder, the Hurricane was slithered sideways to stop off the runway on the frosty grass overshoot area.

There was no further damage and the flight sergeant passed some crisp advice to the Warmwell station armourers about how to fasten gun panels. But this sortie had given further confidence to the theory that fighters could be used in offensive operations at night and that ground targets could be identified and attacked in good moonlight.

The short range of 0·303in machine guns and low lethality meant that these operations were of questionable value apart from waking up the enemy; but we had at least proved that with heavier armament it could be a different story.

A new diversion

Meanwhile a new diversion occurred for 87 Sqn, whose indefatigable CO caused a 450yd strip to be marked out on Charmy Down. He told each flight commander to select four pilots and train them to operate their Hurricanes safely and in bad weather within these take-off and landing limits.

With practice this was just within the Hurricanes' capabilities, and so on May 20 five aircraft flown by Gleed, Ward, Badger, Thoroughgood and the writer set off under low cloud and rain for St Mary's on the Scilly Isles.

The writer had never seen the islands before and, circling in the rain over Hugh Town, watched Gleed's leading aircraft with interest as it slanted down towards

what looked like an impossibly small field on the edge of the cliffs above the town.

The Hurricane dragged low over the rocks, touched short on the grass and slithered down to stop, impressively close to a line of short, wind-bent trees which marked the aerodrome's north-west boundary. Then it was the writer's turn, and with power set to "trickle" in at 85 m.p.h., the wheels were lifted over the rocky cliff edge and planted firmly with chopped throttle on to the grass. Only then did the steepness of the downslope towards the trees become apparent, but the last of the heavily braked energy was killed by kicking on left rudder and skidding the Hurricane to a stop with the wingtip almost in the trees.

Then came the other two aircraft, also with no room to spare, and after instructing the ground party that "Readiness" next day would be from first light, Gleed prepared to go down to the town to find our billets.

At this point a red flare appeared from the direction of the Coastguard station, followed by another, and this was the agreed coded signal to indicate "Enemy aircraft in sight north-east".

Of the four Hurricanes, three were being refuelled and serviced with panels off and only Badger's was still available. He leapt into his cockpit, started up and in a cloud of exhaust smoke, twigs and bits of grass charged back up the slope downwind and disappeared from sight over the cliffs. For an awful moment it seemed that he might have gone in, but then came a clear,

Sqn Ldr Terry Lovell-Grieg, Commanding Officer of 87 Squadron, who was killed in August 1940.

sustained burst of multiple machine-gun fire, a pause and then another burst. In astonished silence we waited for a few minutes and then, with a sudden rumble of sound, the Hurricane appeared low over the headland, swept round the airfield and landed bumpily down the slope. This time pilots and groundcrews were organised in groups to meet it and catch hold of the wingtips to help prevent over-running. Blackened gunports showed evidence of combat.

Sliding back his canopy as the Merlin spluttered to silence, a broadly grinning Peter Badger reported to the CO, "There were two Dornier floatplanes just below the headland—I nearly fell onto one of them as I cleared the cliffs". He had shot one into the sea, and the other had got away in the low cloud.

This was a good start to Operation *Fishing* as the St Mary's detachment was code-named, and there was some celebrating that night in Hugh Town. But this did not prevent Gleed and Thoroughgood coming to "Readiness" at dawn next day on the aerodrome, still in low cloud and rain; and the rest of the pilots were astonished while at breakfast in their Hugh Town boarding house to hear a sudden roar of engines and then many bursts of machine-gun fire.

The "Readiness Section" had scrambled to investigate coastguard flares again and had chased a three-engined Dornier Do 24 flying-boat right across St Mary's at a few hundred feet. They had shot it down just off the coast.

So began a welcome diversion for the 87 Sqn "night fighters", who continued the detached flight at St Mary's, with frequent successes against enemy reconnaissance and shipping raiders and minelayers, for the next six months.

Subsequently the confidence gained in the 1940-41 winter of night and bad weather operations opened the field for wider and more varied operations for fighters in the future. When, in 1942, the then new Hawker Typhoon was getting itself a bad name as a high altitude fighter and was under consideration for cancellation, the writer, as CO of No 609 (WR) Sqn, was authorised to try it out in as many varied roles as was thought practical (see *Typhoon Trials*, August & September 1983 *Aeroplane*). Drawing on 87 Sqn experiences, these trials soon included night exercises of interception and formation, daylight ground-attack in bad weather, moonlight ground-attack on rail targets and shipping attacks by day and night; and in all these the heavy 20mm cannon armament provided the punch to cause real damage up to a useful 1,000yd range, in contrast to the relatively puny 0·303in guns of the Hurricane I.

In a two-month period from December 1942, the Typhoons of 609 Sqn made over 100 successful attacks on trains by moonlight, in addition to achieving a mounting pressure of daylight low-level attacks on rail, canal, transport and airfield targets. The Typhoon was reprieved in 1943 with its role changed from day-interceptor fighter to that of the standard low-level ground-attack fighter of the RAF for the coming invasion of Europe, in which it provided a most successful and often vital contribution.

PROBE PROBARE

Despite enthusiastic efforts over the years by devoted and dedicated supporters of the Hurricane, it has not really been given the credit it deserves as a hard-slogging and tough fighter. A fighter it really was and, although it may have lacked some of the sleekness of its companions and rivals on both sides in the Second World War, it carried as big a punch and was in many respects more amenable to the rough and tumble of a scrap.

Hawker Aircraft had for many years been building, in large numbers, day bombers and interceptor fighters for the Royal Air Force to a fairly standardised type of internal construction and general

In Part 18 of their series on aircraft which received special attention from the Aeroplane and Armament Experimental Establishment, ALEC LUMSDEN and TERRY HEFFERNAN examine the Hawker Hurricane fighter

appearance. Sydney Camm was Chief Engineer at Kingston and, in 1933, was toying with an idea which was to become known as the Fury Monoplane. Basically, this involved mounting the Fury fuselage on a low monoplane wing of Clark YH section, deep, strong and easy to build.

Nowadays that wing might be described as draggy by comparison with the rather more slippery section adopted by Mitchell for his Supermarine fighter, later to be named Spitfire.

Rolls-Royce were working on the P.V.12 engine, later to be known as the Merlin, and Camm decided that the time was right to combine the new fighter with the new engine. Few men could have been better qualified to make such a judgement. The Air Ministry was unusually quick to take the idea up and to issue a specification, F.5/34, round the project, such were the unconstricted but interdependent relations between the leaders of the aircraft industry and the Air Minis-

Flight

Heading photograph, *Flt Lt P. W. S. Bulman, complete with trilby, flying the Hurricane prototype in March 1937.* **Above,** *K5083 landing at Brooklands after a test flight in December 1935. The Hawker Monoplane F.36/34, as it was still known, was sent to A&AEE at Martlesham Heath three months later.*

try. The specification was revised to F.36/34, and from this emerged the Hawker Hurricane.

How the Hurricane acquired its formidable battery of eight guns, rather than the then customary pair of fuselage-mounted guns, has been told many times in various ways. The Air Ministry's experts, notably the then Sqn Ldr R. S. Sorley (to become Air Marshal Sir Ralph) had worked out, following the triumphant winning of the Schneider Trophy and the raising of the world speed record to over 407 m.p.h., that one swift pass and a two-second burst of fire was all that a fighter pilot was likely to be able to achieve in intercepting a bomber. To destroy a target in that time required the combined fire of eight machine guns, each firing 1,000 rounds per minute. Just such a gun in the shape of the 0·303-in Browning became available at the crucial moment and, thanks to its thick wing, the Hurricane was able to accommodate four guns and their ammunition tanks on each side, mounted in a close group.

The Prototype

Hawker F.36/34 K5083 was powered by the Merlin C which gave 1,029 h.p. to a big wooden Watts propeller. It was taken by road from Kingston to Brooklands where P. W. S. Bulman flew it on November 6, 1935. The first reports on its behaviour were good and it was only four months before the "Hawker Monoplane F.36/34, K5083", as it was still called, was sent to Martlesham Heath, the home near Ipswich in Suffolk of the Aeroplane and Armament Experimental Establishment, for initial Service testing.

Left, *the prototype Hurricane photographed during an engine run at Brooklands in the winter of 1935–36.* **Right,** *the same aircraft being prepared for a test flight by Flt Lt Bulman at Brooklands in December 1935.*

The first series of tests on K5083 at Martlesham was between February and April, 1936, a report made to the Air Ministry being drafted at the end of April. Handling trials were made with the aircraft loaded to 5,672lb all-up weight.

The aileron controls operated freely and without play when the aircraft was on the ground and the pilot could obtain full sideways movement without restriction. In the air, the ailerons were light at low speeds when climbing and on the glide. A steady increase in heaviness resulted from increases in speed. At maximum level speed and in the dive control was found to be heavy for a fighter. A small peculiarity was that, at moderate speeds only, when the starboard aileron was raised, the feel suddenly became lighter and the control more effective. However, the response to the ailerons was rapid and they were effective under all normal manœuvres in flight. During take-off, landing and at the stall, the response was less rapid although remaining satisfactory. The aileron control was also generally satisfactory, although if it could be made lighter at high

Flight

speeds without over-balancing, it would be improved.

The rudder control was similar to the ailerons in response, light on the ground and at low speeds but increasing in heaviness with increase in speed and at high speeds extremely heavy. All the same, response to rudder pressure was quick and effective as far as it went in all conditions of flight, although it would have been better if made lighter over 150 m.p.h. The bias control gear (trimmer) was quick and easy to operate once the trimming strip had been correctly adjusted.

The elevator control was satisfactory, being light, effective and giving a quick response under all conditions from the stall to diving speeds. The tail trimming gear was easy and light in operation although there was not quite enough range of controls to trim the aircraft in all flight conditions. A slight increase in the range at both ends of the scale might be necessary to allow for changes in the c.g. position.

Although there was no slippage in the operating cables, they were inclined to

Flight

stretch, allowing some free movement in the tabs and giving an unpleasant effect of fore and aft instability. The trim settings found to be effective were: fully forward on the climb; not quite fully forward at maximum and cruising speeds; fully back on approach.

As for other controls, those for the engine were well placed and satisfactory; the flap control was well placed, easy to operate and took 10 to 15 seconds to move over the full range. There was a noticeable change of trim with flaps down although the trimming gear did not need to be adjusted until they were fully down, thanks to the effectiveness of the elevators. The flaps steepened the glide and were very effective, even improving the aileron control when down. The brakes were smooth, progressive and easy to operate although, since they became progressively more effective towards the end of the landing run, care had to be exercised to prevent the aircraft from tipping onto its nose.

Laterally the F.36/34 was stable, tending to fly left wing low on the climb and right wing low at top speed. Directionally it was stable under all conditions. It was neutrally stable longitudinally, with engine on or off, as loaded for the tests. The stall was normal with no vice or snatching at the controls. The aircraft handled well in the aerobatics which were tested—loops, half-rolls off the top and stalled turns. At moderate speeds it handled well but, at high speeds, the aerobatic handling would have been improved by lighter rudder and ailerons. The F.36/34 was easy and normal to take off and to land. It had a tendency to swing to the left when taking off but that could easily be held by the rudder and there was no swing when landing. If the throttle was opened with undercarriage and flaps down, trimmed for landing, the aircraft could be held by the elevator control. It was difficult to sideslip and could not be held beyond about 10°. Ground handling was easy and stable in winds up to 30 m.p.h.

The inwards-retracting undercarriage was found to be very satisfactory, having good shock absorbing qualities and rebound damping. The retracting gear (operating all three wheels) was simple and easy to operate; worked by hand-pump, the wheels could be retracted in about 45sec without undue exertion by the pilot and could be lowered in about 20sec. The indicator worked satisfactorily and the wheels themselves could be seen when up or down through small windows in the floor, the latter being described as an excellent feature. No mention was made of the effect on the view through these windows when they were covered in oil but, no doubt, such things were not allowed to occur at Martlesham.

The view forwards and around the upper hemisphere was good except for the blind spot directly aft of the pilot's head, which obscured the tail. The comment was made that "In a fighting aircraft, view in

this direction may be important, though for Home Defence purposes less important, and if this blind spot could be eliminated, the fighting view would be much improved". The preoccupation with bomber interception is clearly evident here. The view downwards was largely blanked by the wings but for take-off and landing was good. "The covered cockpit enables the pilot to look aft without the risk of having his goggles blown off."

The A&AEE reported that the cockpit was extremely roomy and comfortable as well as keeping warm down to an outside air temperature of -50°C. It was not unduly noisy and the layout of the controls was satisfactory. The cockpit was easy to enter and leave when on the ground with the hood fully open but it was found that at speeds above about 150 m.p.h. it was impossible to slide the cockpit roof to the open position. At such speeds the air pressure would slide the hood from the open to the closed position. Consequently, it was quite impossible for the pilot to make an emergency exit at any speed above 150 m.p.h., "a defect unacceptable in a high speed fighting

aircraft". Modification was required to the cockpit hood so that it could be operated at any speed.

In the light of the above, it is characteristic of the A&AEE pilots (and indeed of the profession as a whole) that diving tests were made to a shade over the prototype's limiting speed of 300 m.p.h. or 3,150 r.p.m. (whichever occurred first). The dives were entered from trimmed level flight conditions. From 14,000ft, the coarse-pitch wooden propeller maintained 2,700–2,800 r.p.m. with throttle settings between fully open and 1/3 open at speeds of up to 310 m.p.h. The aircraft was steady in dives, and small movements of the controls led to the correct response without any signs of control surface instability or vibration. Recovery was easily effected in all cases although the controls became very heavy.

No trouble was experienced with the airframe of the Hawker Monoplane while at Martlesham, with the exception of damage to the port undercarriage due to the lack of clearance of the hinged fairing. This was caused by the fairing striking a tuft of earth and grass roots on landing.

Flight

Left, *this close-up photograph of the prototype Hurricane shows well the inward retracting undercarriage, the big wooden Watts propeller and the neat cowling of the early Model C Merlin engine.*

Above, *Hurricane I L1582 was part of the first batch of 600 Hurricanes delivered to the RAF between December 1937 and November 1939. It served with Nos 3 and 73 Squadrons but on August 4, 1939 was lost in a crash landing at RAF Digby.*

Right, *first of the initial production batch was Hurricane I L1547, first used by the A&AEE for spinning trials. This Hurricane was lost in the Mersey off Ellesmere Port on October 10, 1940.*

The only other damage was to the brake pipeline and it was recommended that the fairing be modified to give more clearance from the ground, and that the pipe be attached to the leg rather than the fairing. The hinged fairing was soon to be removed, as were the two temporary tailplane struts.

A certain amount of engine trouble occurred to the early model Merlin C, No 15, which was fitted to K5083. An inspection revealed the failure of supercharger bearings and the engine was changed. Further engine trouble revealed a piston failure. Another engine change to Merlin No 19 resulted in still more trouble and temporary modifications being made, including carburation, magnetos and sparking plugs. Following a 20-hour strip down, three broken valve springs were found. Two of the failures which occurred were due to faulty automatic boost control capsules. Several engine cuts occurred when Service pilots were flying and each occurred after 1hr 10min. It was subsequently found that the port tank was empty on each occasion, the aircraft having been flown with all tanks "on",

causing vapour locks in the fuel system. The carburation was too sensitive to slight mixture adjustments to be suitable for Service use. The engine was, however, not representative of the type to be used in production aircraft.

The Hawker F.36/34 was described as simple and easy to fly, with no apparent vices. The controls were satisfactory except at high speeds. Take-off and landing presented no difficulties nor, despite the high top speed, did the approach and landing with wheels and flaps down. Curiously, no mention was made of spinning.

Hurricane I

In June, 1936, the name Hurricane having been given to the Hawker F.36/34, a production order for 600 was placed with Hawker Aircraft. The first production aircraft did not fly until October 12, 1937, due largely to a decision to use the Merlin II rather than the Mk I. This decision had resulted in a number of small but important modifications to the engine installation including cowlings. Build-up of the production Hurricane I was rapid, thanks to the simplicity of its construc-

tion. By Christmas 1937, 111 Sqn at Northolt was already exchanging its Gauntlets for the new monoplanes.

Rapid production and entry into squadron use allowed experience with the Hurricane I to be gained quickly, enabling faults arising in service to be rectified. An Air Ministry letter dated February 14, 1938 required tests to be made following trouble with guns failing to operate satisfactorily in the intense cold at high altitude. The heating of the gun bays (by means of air scoops behind the radiator) required improvement, as did the compressed air operating system.

The first Hurricane Is to enter squadron service had rear ends similar in shape to the F.36/34 prototype. Spinning had not been cleared satisfactorily owing to the delays in the production programme and it soon became clear that extra keel area aft was needed.

An underfin was added, the rudder was deepened to give extra area and the tailwheel was no longer retractable. Trials were made at Martlesham between September 2, 1938 and January 17, 1939 involving the modified first production

Hurricanes of the Northolt-based 111 Squadron

aircraft, L1547 and L1696. For this series of tests, both had wooden airscrews and "streamline" rather than ejector exhausts. Hurricane L1547 was used for spinning and the other for diving trials.

Three pilots carried out spins on L1547 and all spins were entered from straight stalls with wheels and flaps up. Spins were normal, generally smooth after the first one or two turns, and recovery was complete within one or two turns after the correct sequence of control movements. These were: full opposite rudder, and stick eased slowly forward to a central position. Level flight was regained after a height loss of about 3,800ft in a three-turn spin. Too quick or coarse elevator movement resulted in another stall and a flick the opposite way. Too much forward movement of the stick resulted in an unacceptable height loss. In March, 1939, L1547 was also used for spinning tests with a two-pitch D.H. airscrew, in which the results were similar, being described as "easy, smooth and pleasant" after the irregularity of the first two or three turns.

Diving trials on L1696 confirmed the recommendations following the prototype trials. The aircraft was dived beyond its limiting speed of 380 m.p.h. IAS and although the controls became progressively heavier, they were at no time excessively so. Small changes of longitudinal trim through slight rudder movements were easily held by the stick. Variations of c.g. affected neither control loads nor recovery but the radiator flap when open caused progressive tail-heaviness with increase of speed. The hood was opened to about half-way from the closed position in dives to about 380 m.p.h. The elbow would be needed to open it further, with serious risk to the arm in the slipstream in the event of slipping. The half-way position did not engage the open lock, nor did it allow the emergency panel on the starboard side to be opened. Mechanical means for opening the hood fully were recommended.

Resulting from night flying trials, involving three Hurricanes, L1547, L1574 and L1696, the streamline exhausts which gave a little less glare were slightly preferred to ejectors. The landing lights were satisfactory as was the Hurricane's handling (it may be recalled that, subsequent to the adoption of ejector exhausts, "blinkers" were mounted on the fuselage sides to shield the pilot's eyes).

A detailed handling report on the Hurricane I with a wooden propeller was submitted to the Air Ministry on April 6, 1939. Although detail improvements had been made following the prototype's visit to Martlesham, Sydney Camm had got the design about right from the start. The flaps and undercarriage were operated hydraulically, a joint selector lever in an H-shaped gate being coupled to a power lever in similar fashion to the Henley. Flap and undercarriage operations and position indicators were satisfactory. The throttle friction device could be slackened off too much, allowing the throttle to creep back, perhaps at a crucial moment, when the left hand was removed to raise the undercarriage and flaps.

Brakes, instruments and seat adjustment were satisfactory. Cockpit lights and, in particular, the undercarriage lights required adjustment, the latter being too bright and needing to be switched off when landing at night. Handling on the ground and in the air was satisfactory, control response generally being light and sensitive. The Hurricane tended to swing left at speeds up to 280 m.p.h. IAS and to the right at higher speeds. It stalled at 73 m.p.h., flaps and wheels up and 62 m.p.h. "all down", more gently when clean than in the latter case and always dropping the port wing. Although there were longitudinal oscillations when near the stall, there was little warning or buffeting. Turns could be made satisfactorily at 1.1 times the stalling speed. The aeroplane was easy to land, final approach speed being 80 m.p.h. At 75 m.p.h., it stalled before the stick came right back, the left wing dropping. Handling in aerobatics was normal, although, if speed was allowed to drop in the nearly vertical position during a loop, a "forced stall" would occur, the aeroplane flicking off its back. This was described as "a little disturbing".

A series of handling tests was made from November 7–23, 1938 on L1547 with D.H. two-position variable-pitch propellers and ejector exhausts. The extra weight of the VP propeller made the Hurricane I nose-heavy in the glide with flaps down, even with full tail-down trim. On the ground it required care in braking, especially on rough ground.

Although all marks of the great Hawker fighter demanded respect when flown at low speeds and, for a careless pilot, could be very unforgiving, overall the handling was pleasant; the aircraft well deserved the title of "a gentleman's aerial carriage". Soon after these tests, the A&AEE went to Boscombe and the Hurricane I went to war. The rest is history.

Hawker F.36/34 data

Engine: Rolls-Royce Merlin C, 1,029 b.h.p. at 2,600 r.p.m. at rated pressure +6 lb/in², at 11,000ft.

Dimensions

Span	40ft 0in
Length	31ft 6in
Height (propeller vertical)	13ft 6in
Wing area	258ft²

Weights

Tare	4,129lb
Loaded	5,672lb

Performance

Take-off run (wind 5 m.p.h.)	265yd
Take-off run to clear 50ft	430yd
Rate of climb, sea level	2,550ft/min
Climb to rated altitude	4min
Stalling, flaps up	77 m.p.h.
Stalling, flaps down	57 m.p.h.
Landing run, flaps & brakes	205yd
Max level speed, 16,000ft, 2,950 r.p.m.	315 m.p.h.
Service ceiling (estimated)	34,000ft

De Havilland
MOSQUITO
PORTFOLIO

Handley Page
HALIFAX
PORTFOLIO